Foundations of Control Engineering

Foundations of Control Engineering

Marc Bodson

Front cover: space shuttle robotic arm (see p. 38, credit: NASA).
Back cover: Watt's governor (see p. 2, photo by the author), X-29 (see p. 3, credit: NASA), closed-loop frequency response (see p. 158, graphic by the author).

Preface

The book presents the core theory of control engineering, together with its foundations in signals and systems. These foundations include continuous-time systems using the Laplace transform, discrete-time systems using the z-transform, and sampled-data systems connecting the two domains. The classical theory of control covers the analysis of the dynamic response of linear time-invariant systems, root-locus techniques for feedback design, and the frequency-domain analysis of closed-loop systems. Control engineering is strongly related to signal processing and communications, and the book includes a discussion of phase-locked loops as an example of feedback control. To the extent possible, the origin of the theoretical results is explained, and the technical details needed to reach a more complete understanding of the concepts are included. On the other hand, the book does not present design studies or specialized topics, for which the reader is referred to the bibliography. Material complementing the book is available through the author's web page, including solutions to selected problems and virtual lab experiments.

About the author

Marc Bodson received a Ph.D. degree in Electrical Engineering and Computer Science from the University of California, Berkeley, in 1986. He obtained two M.S. degrees - one in Electrical Engineering and Computer Science and the other in Aeronautics and Astronautics - from the Massachusetts Institute of Technology, Cambridge MA, in 1982. In 1980, he received the degree of Ingénieur Civil Mécanicien et Electricien from the Université Libre de Bruxelles, Belgium. He is a Professor of Electrical & Computer Engineering at the University of Utah in Salt Lake City, where he was Chair of the department between 2003 and 2009. He was the Editor-in-Chief of *IEEE Trans. on Control Systems Technology* from 2000 to 2003. He was elected *Fellow* of the IEEE in 2006, and *Associate Fellow* of the American Institute of Aeronautics and Astronautics in 2013. He also received the Engineering Educator of the Year award from the Utah Engineers Council in 2007. His activities are described in further detail at www.ece.utah.edu/~bodson.

Contents

Chapter 1

Introduction to feedback systems

1.1 Standard feedback system

Fig. 1.1 shows a standard feedback system. The elements of the system are: P, the plant, C the compensator, and $\oplus$, a summing function or comparator. The signals are r, the reference input, $e = r - y$, the tracking error, u, the control input, and y, the plant output. The main objective of such a feedback system is to achieve tracking of the reference input by the plant output, *i.e.*, to make the tracking error as close to zero as possible. Examples of feedback systems encountered in everyday life include the thermostat that regulates the temperature of a room, and the cruise control that regulates the speed of a car.

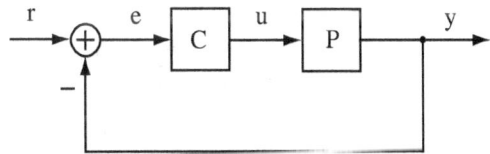

Figure 1.1: Standard feedback system

In addition to tracking the reference input, it is important to have an acceptable transient response when the reference input changes. Fig. 1.2 shows three examples of transient responses. Response (a) is a satisfactory response, while responses (b) and (c) are undesirable. Response (b) is fast at first, but converges to the steady-state very slowly. Response (c) overshoots the steady-state and oscillates multiple times around the desired value. Challenges in achieving good steady-state and transient responses in control systems include:

- errors in the measurements (noise and offsets),

- disturbances affecting the system (such as the slope of the road in a cruise control system),

- uncertainty about the plant characteristics (including changes over time),

- limitations of the actuation system (delays and finite range of the control input).

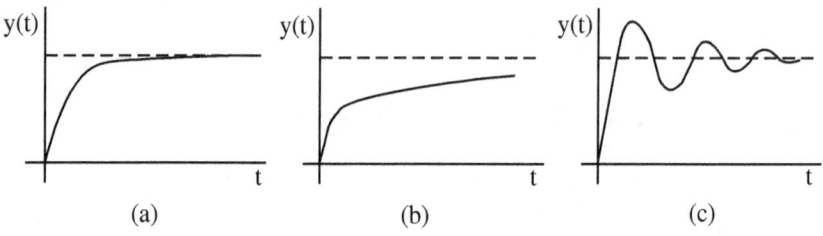

Figure 1.2: Examples of transient responses

1.2   Example of Watt's governor

Watt's governor, shown in Fig. 1.3, is an early example of control system originating from the industrial revolution. The control input is the flow of steam applied to an engine, and the plant output is the rotational speed. Through the centrifugal force acting on two spheres, the speed of rotation creates a negative feedback that opens or closes a valve feeding steam to the engine. In this manner, speed was regulated to some desired set point. An interesting paper [20] by J. C. Maxwell (author of Maxwell's equations) describes early attempts to analyze governors as linear dynamical systems, and to design mechanical feedback systems to achieve specific control objectives. Much progress was achieved over time in the understanding of these concepts.

1.3   Example of flight control system

A more modern example is the flight control system for an advanced aircraft, shown schematically in Fig. 1.4. The reference input is provided by the pilot, in the form of commands in the pitch, roll, and yaw axes (given by stick and pedal movements). The flight control system uses the pilot commands in combination

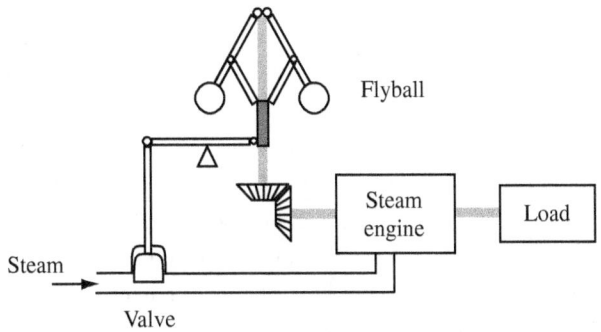

Figure 1.3: Watt's governor

with measurements of the aircraft states (for example, its angular velocities) to determine the appropriate commands to be applied to the control surfaces.

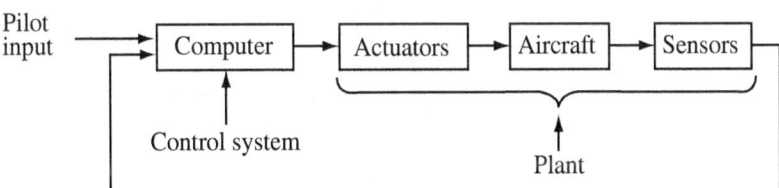

Figure 1.4: Flight control system

In some modern aircraft, such as the X-29, the dynamic behavior is so unstable that a pilot would be unable to maintain steady flight without the feedback actions implemented by the flight control computer. In the worst flight condition of the X-29, for example, an angular deviation from horizontal flight doubled every 0.12 seconds (the stabilization task is equivalent to the one required to balance a 17.4 in stick on a finger) [7]. Computations were performed at a rate of 40 times per second to provide adequate stabilization and control of the aircraft.

1.4 Example of active noise control

Fig. 1.5 shows a different example of feedback system, specifically an active noise control system. Here, the purpose of the control system is to make a region of space free from noise. The result is achieved by generating a sound wave using a speaker and canceling exactly the wave produced by the noise source. The air pressure produced by the speaker and by the noise source at the microphone is shown as a function of time in Fig. 1.6. Ideally, the two waves will cancel each other exactly. The control problem is quite different from the flight control application. There is no issue of stabilization of the plant, or of tracking of commands. The problem is purely one of disturbance rejection. Challenges in this application are the speed at which computations must be performed (a rate of 8 kHz is typical), and the time delay present in the plant (due to the time it takes for the sound to travel from the speaker to the microphone). The general structure of the control system, however, is similar to the structure of the flight control system of Fig. 1.4, where the actuators are replaced by the speaker, the sensors by the microphone, the aircraft by the acoustics from the speaker to the microphone, and the reference input is set to zero.

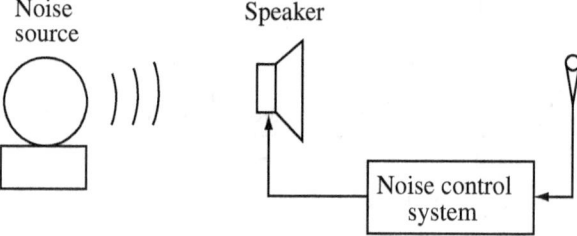

Figure 1.5: Active noise control system

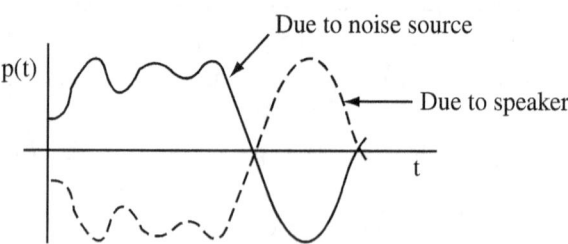

Figure 1.6: Principle of noise cancellation

Chapter 2

Continuous-time signals

2.1 The Laplace transform

2.1.1 Definition

A *continuous-time signal* is a function of a variable t called time, which is defined over the set of real numbers. The Laplace transform of a signal $x(t)$ is given by

$$X(s) = \int_0^\infty x(t)e^{-st}dt. \tag{2.1}$$

The Laplace transform $X(s)$ is a real function of the complex variable s.

2.1.2 Examples

1) $x(t) = \delta(t) \Leftrightarrow X(s) = 1$,
   where $\delta(t)$ is an *impulse function* or Dirac function.

2) $x(t) = u(t) \Leftrightarrow X(s) = \frac{1}{s}$,
   where $u(t)$ is a *step function*: $u(t) = 0$ for $t < 0$ and $u(t) = 1$ for $t \geqslant 0$.

3) $x(t) = 1 \Leftrightarrow X(s) = \frac{1}{s}$.
   The transform is the same as for $u(t)$, because $X(s)$ does not depend on $x(t)$ for $t < 0$.

4) $x(t) = e^{at} \Leftrightarrow X(s) = \frac{1}{s-a}$.

5) $x(t) = \cos(bt) \Leftrightarrow X(s) = \frac{s}{s^2 + b^2}$.

6) $x(t) = \sin(bt) \Leftrightarrow X(s) = \frac{b}{s^2 + b^2}$.

7) $x(t) = t \Leftrightarrow X(s) = \frac{1}{s^2}$.

8) $x(t) = t^n e^{at} \Leftrightarrow X(s) = \frac{n!}{(s-a)^{n+1}}$.

9) $x(t) = e^{at} \cos(bt) \Leftrightarrow X(s) = \dfrac{s-a}{(s-a)^2 + b^2}$.

10) $x(t) = e^{at} \sin(bt) \Leftrightarrow X(s) = \dfrac{b}{(s-a)^2 + b^2}$.

11) $x(t) = te^{at} \cos(bt) \Leftrightarrow X(s) = \dfrac{(s-a)^2 - b^2}{((s-a)^2 + b^2)^2}$.

12) $x(t) = te^{at} \sin(bt) \Leftrightarrow X(s) = \dfrac{2b(s-a)}{((s-a)^2 + b^2)^2}$.

Many transforms can be obtained from the formula for $t^n e^{at}$. For example,

$$x(t) = te^{at} \cos(bt) = te^{at} \left(\frac{e^{jbt} + e^{-jbt}}{2} \right) = \frac{t}{2} e^{(a+jb)t} + \frac{t}{2} e^{(a-jb)t}, \tag{2.2}$$

so that

$$
\begin{aligned}
X(s) &= \frac{1/2}{(s-a-jb)^2} + \frac{1/2}{(s-a+jb)^2} = \frac{(s-a+jb)^2/2 + (s-a-jb)^2/2}{((s-a)^2 + b^2)^2} \\
&= \frac{(s-a)^2 + b^2}{((s-a)^2 + b^2)^2}. \tag{2.3}
\end{aligned}
$$

Although the intermediate steps use complex variables, the final result is a real function of the variable s, which must be the case when the signal is real.

2.1.3   Relationship between pole locations and signal shapes

All the examples of the previous section correspond to transforms of the form $X(s) = N(s)/D(s)$, where $N(s)$ and $D(s)$ are polynomials in s. Such functions are called *rational functions* of s. The coefficients of the polynomials are real, so that the roots of the numerator and denominator polynomials are real, or occur in complex pairs. The roots of $D(s)$ are called the *poles* of $X(s)$, while the roots of $N(s)$ are called the *zeros* of $X(s)$. The examples show that there is a direct relationship between the location of the poles and the shapes of the signals. The signals are of the form $t^n e^{at} \cos(bt + \phi)$, with the coefficient of the exponential a and the frequency of the sinusoid b corresponding to the real part and imaginary part of the pole (respectively), and n (the power of t) being equal to the multiplicity of the pole minus 1. Fig. 2.1 illustrates the relationship between pole locations and signal shapes in the case of single poles and Fig 2.2 does the same for double poles.

   In general, the location of the poles along the real axis determines the rate of growth or decay of the signal. For a pole in the left half-plane (with $\mathrm{Re}(p) =$

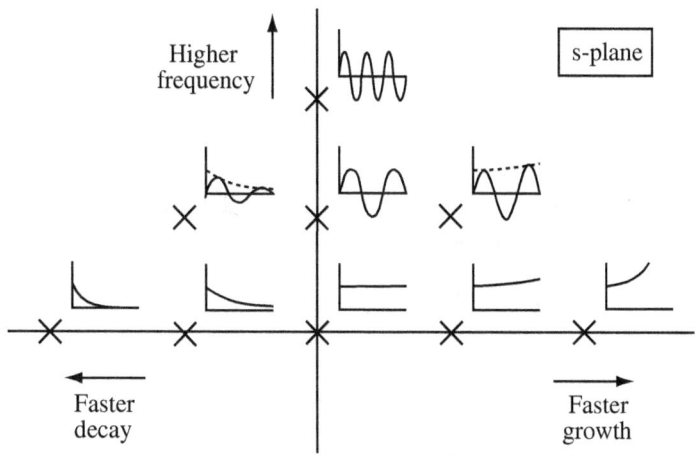

Figure 2.1: Signal shape *vs.* pole location (single pole)

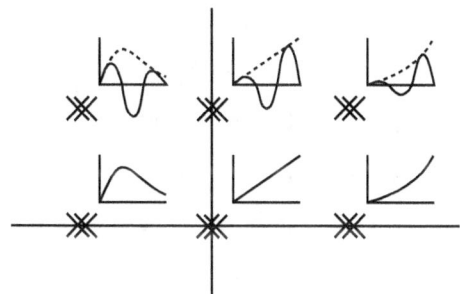

Figure 2.2: Signal shape *vs.* pole location (double pole)

$a < 0$), the signal contains a decaying exponential with

$$
\begin{aligned}
e^{at} &= 0.37 \quad \text{for} \quad t = -\frac{1}{a} \quad (63\% \text{ convergence}). \\
&= 0.14 \quad \text{for} \quad t = -\frac{2}{a} \quad (86\% \text{ convergence}). \\
&= 0.05 \quad \text{for} \quad t = -\frac{3}{a} \quad (95\% \text{ convergence}). \\
&= 0.02 \quad \text{for} \quad t = -\frac{4}{a} \quad (98\% \text{ convergence}). \quad (2.4)
\end{aligned}
$$

The value of time $\tau_c = -1/a$ is usually called the *time constant*. The value $\tau_{2\%} = 4\tau_c$ corresponding to a 2% residual is often taken to be the *convergence time*, or *settling time*. For a growing exponential (or pole in the right half-plane)

$$e^{at} = 2.0 \qquad \text{for} \qquad t = \frac{0.7}{a} \tag{2.5}$$

and the value of the time is called the *time to double* the amplitude. For $a = 10$ rad/s, $\tau_{double} = 70$ ms.

The imaginary part of the pole b gives an indication of the frequency of oscillation. Specifically, $\cos(bt)$ has a period

$$T_{osc} = \frac{2\pi}{b} \qquad \text{(period of oscillation)}. \tag{2.6}$$

For $b = 100$ rad/s, $T_{osc} = 62.8$ ms.

2.1.4   Properties of the Laplace transform

1) *Linearity:* $y(t) = a_1 x_1(t) + a_2 x_2(t) \Leftrightarrow Y(s) = a_1 X_1(s) + a_2 X_2(s)$.

2) *Differentiation:* $y(t) = \frac{dx(t)}{dt} \Leftrightarrow Y(s) = sX(s) - x(0)$
   (use $x(0^-)$ for discontinuous signals).

3) *Integration:* $x(t) = x(0) + \int_0^t y(\sigma)d\sigma \Leftrightarrow X(s) = \frac{x(0)}{s} + \frac{Y(s)}{s}$.

4) *Final value:* if $\lim_{t\to\infty} x(t)$ exists, $\lim_{t\to\infty} x(t) = \lim_{s\to 0} sX(s)$.

5) *Multiplication by t:* $y(t) = tx(t) \Leftrightarrow Y(s) = -\frac{dX(s)}{ds}$.

6) *Multiplication by e^{at}:* $y(t) = e^{at}x(t) \Leftrightarrow Y(s) = X(s - a)$.

7) *Delay (right shift):* $y(t) = x(t - T)u(t - T)$ for $T > 0 \Rightarrow Y(s) = X(s)e^{-sT}$.

Comments on the properties

- Linearity is a key property that is used to find the time signals corresponding to rational Laplace transforms using partial fraction expansions.

- In short, differentiation $\Leftrightarrow$ multiplication by s and integration $\Leftrightarrow$ division by s. These properties are very useful for the analysis of ordinary differential equations using the Laplace transform.

- The correct application of the final value theorem is tricky, because one must know that the limit exists. For example, $\frac{1}{s+1}$ and $\frac{1}{s-1}$ yield the same result, but the limit exists only in the first case. For certain transforms, the existence of the limit can be determined from $X(s)$, as will be seen later.

- For the delay property, one must have that $T > 0$, *i.e*, that the signal is shifted to the right, or delayed. The property does not apply for a shift to the left.

2.2 Inverse of Laplace transforms using partial fraction expansions

2.2.1 General form of a partial fraction expansion

In this section, we restrict our attention to Laplace transforms that are rational functions of s. The procedure of partial fraction expansions allows one to break-up the transforms into simpler terms, which may be transformed back into the time domain using standard formulas.

Fact - Partial fraction expansion
1. Let $X(s) = N(s)/D(s)$, where $N(s)$ and $D(s)$ are polynomials in s with real coefficients, no common roots, and $\deg N(s) < \deg D(s)$. Let p_1, p_2, ..., p_n be the roots of $D(s) = 0$, with multiplicities r_1, r_2, ..., r_n, respectively, so that $D(s) = (s - p_1)^{r_1}(s - p_2)^{r_2}...(s - p_n)^{r_n}$.
Then, $X(s)$ can be expanded as

$$X(s) = \sum_{i=1}^{n}\sum_{k=1}^{r_i} \frac{c_{i,k}}{(s - p_i)^k}. \tag{2.7}$$

Note that the number of terms is equal to the degree of $D(s)$. The coefficients $c_{i,k}$ may be complex, but the coefficients associated with complex conjugate poles are themselves complex conjugate: $p_l = p_m^* \Rightarrow c_{l,k} = c_{m,k}^*$.
2. The corresponding signal $x(t)$ is given by

$$x(t) = \sum_{i=1}^{n}\sum_{k=1}^{r_i} c_{i,k}\frac{t^{k-1}}{(k-1)!}e^{p_i t}, \tag{2.8}$$

where $0! = 1$. Although the coefficients and the functions may be complex, complex conjugate terms can be grouped to yield real functions, as will be shown later.

2.2.2 Determination of the coefficients

Two methods exist for the determination of the coefficients: a direct method called *residue method*, or *cover-up method*, and an indirect method called *method of clearing fractions*. Examples are discussed later to illustrate the application of the methods.

Residue method (or cover-up method)

The coefficients are given by the following formulas

$$c_{i,r_i} = \left[(s - p_i)^{r_i} X(s)\right]_{s=p_i},$$

$$c_{i,r_i-1} = \left[\frac{d}{ds}(s - p_i)^{r_i} X(s)\right]_{s=p_i},$$

$$c_{i,r_i-m} = \frac{1}{m!}\left[\frac{d^m}{ds^m}(s - p_i)^{r_i} X(s)\right]_{s=p_i}, \qquad (2.9)$$

where the notation $[\cdot]_{s=p_i}$ means that the expression inside the bracket is evaluated at $s = p_i$.

Clearing fractions

Equating the numerators, so that $X(s) = N(s)/D(s)$, one finds that

$$\begin{aligned}
N(s) = \ & c_{1,1}(s - p_1)^{r_1-1}(s - p_2)^{r_2}...(s - p_n)^{r_n} \\
& +c_{1,2}(s - p_1)^{r_1-2}(s - p_2)^{r_2}...(s - p_n)^{r_n} \\
& +... + c_{1,r_1}(s - p_2)^{r_2}...(s - p_n)^{r_n} \\
& +c_{2,1}(s - p_1)^{r_1}(s - p_2)^{r_2-1}...(s - p_n)^{r_n} \\
& +c_{2,2}(s - p_1)^{r_1}(s - p_2)^{r_2-2}...(s - p_n)^{r_n} \\
& +... + c_{2,r_2}(s - p_1)^{r_1}(s - p_3)^{r_3}...(s - p_n)^{r_n} \\
& +...
\end{aligned} \qquad (2.10)$$

Matching the coefficients of the polynomials on both sides yields a system of linear equations in the unknowns $c_{i,k}$.

2.2.3 Grouping complex terms

Grouping terms in the time domain

The sum of two terms of the form

$$X(s) = \frac{c}{(s - p)^k} + \frac{c^*}{(s - p^*)^k} \qquad (2.11)$$

gives the time function

$$x(t) = c\frac{t^{k-1}}{(k - 1)!}e^{pt} + c^*\frac{t^{k-1}}{(k - 1)!}e^{p^*t}. \qquad (2.12)$$

Denoting $a = \text{Re}(p)$, $b = \text{Im}(p)$, the function becomes

$$
\begin{aligned}
x(t) &= \frac{t^{k-1}}{(k-1)!} e^{at} \left(ce^{jbt} + c^* e^{-jbt} \right) \\
&= \frac{t^{k-1}}{(k-1)!} e^{at} \left((c + c^*) \cos(bt) + j(c - c^*) \sin(bt) \right), \quad (2.13)
\end{aligned}
$$

which shows that $x(t)$ is the real function of time

$$
x(t) = 2 \,\text{Re}(c) \frac{t^{k-1}}{(k-1)!} e^{at} \cos(bt) - 2 \,\text{Im}(c) \frac{t^{k-1}}{(k-1)!} e^{at} \sin(bt).
$$
$$(2.14)$$

The time function can also be expressed as

$$
x(t) = 2 \,|c| \frac{t^{k-1}}{(k-1)!} e^{at} \cos(bt + \angle c), \quad (2.15)
$$

since

$$
\begin{aligned}
2 \,|c| \cos(bt + \angle c) &= 2 \,|c| \left(\cos(\angle c) \cos(bt) - \sin(\angle c) \sin(bt) \right) \\
&= 2 \,\text{Re}(c) \cos(bt) - 2 \,\text{Im}(c) \sin(bt). \quad (2.16)
\end{aligned}
$$

Important observation: a pole at $s = a + jb$ and its complex conjugate together yield a time function that is the product of three functions: an exponential function whose coefficient is the real part of the pole, a sinusoidal function whose angular frequency is the imaginary part of the pole, and the time function raised to a power equal to the multiplicity of the pole minus one. The residue of a pole determines the scalar by which the function is multiplied, as well as the phase of the sinusoidal function.

Grouping terms in the s-domain

The grouping of complex terms can also be performed in the s-domain. This fact is useful in the procedure of clearing fractions, in order to obtain a system of linear equations with real coefficients. For example, two terms due to non-repeated imaginary poles can be combined as follows:

$$
\begin{aligned}
X(s) &= \frac{c}{s - jb} + \frac{c^*}{s + jb} = \frac{(c + c^*)s + jb(c - c^*)}{(s - jb)(s + jb)} \\
&= \frac{2 \,\text{Re}(c)s - 2 \,\text{Im}(c)b}{s^2 + b^2}. \quad (2.17)
\end{aligned}
$$

The result is identical to the one obtained in the time domain, because $\frac{s}{s^2 + b^2}$ is the transform of $\cos(bt)$ and $\frac{b}{s^2 + b^2}$ is the transform of $\sin(bt)$. In general, any two terms

$$X(s) = \frac{c}{(s-a-jb)^k} + \frac{c^*}{(s-a+jb)^k}$$
$$= \frac{2\operatorname{Re}(c)\operatorname{Re}(s-a+jb)^k - 2\operatorname{Im}(c)\operatorname{Im}(s-a+jb)^k}{((s-a)^2+b^2)^k}, \quad (2.18)$$

where s is treated as a real variable when taking the real and imaginary parts of $(s-a+jb)^k$. Again, the result is identical to the one obtained in the time domain because

$$x_1(t) = \frac{t^{k-1}}{(k-1)!}e^{at}\cos(bt) \quad \Leftrightarrow \quad X_1(s) = \frac{\operatorname{Re}(s-a+jb)^k}{((s-a)^2+b^2)^k},$$
$$x_2(t) = \frac{t^{k-1}}{(k-1)!}e^{at}\sin(bt) \quad \Leftrightarrow \quad X_2(s) = \frac{\operatorname{Im}(s-a+jb)^k}{((s-a)^2+b^2)^k}. \quad (2.19)$$

2.2.4   Examples

(1-a) Example with real poles by residue method

$$X(s) = \frac{1}{(s+1)^2(s+2)} = \frac{c_{11}}{s+1} + \frac{c_{12}}{(s+1)^2} + \frac{c_{21}}{s+2}. \quad (2.20)$$

For c_{21}, the formula gives

$$c_{21} = [(s+2)X(s)]_{s=-2} = \left[\frac{1}{(s+1)^2}\right]_{s=-2} = \frac{1}{(-2+1)^2} = 1. \quad (2.21)$$

Note that the formula can be explained easily in this case by noting that if $X(s)$ is multiplied by $s+2$, one gets

$$(s+2)X(s) = \frac{c_{11}(s+2)}{s+1} + \frac{c_{12}(s+2)}{(s+1)^2} + c_{21}. \quad (2.22)$$

Only c_{21} remains in this expression when $s = -2$.

The parameter c_{12} (the coefficient of the highest power for the pole at $s = -1$) is determined in a similar way

$$c_{12} = [(s+1)^2X(s)]_{s=-1} = \left[\frac{1}{s+2}\right]_{s=-1} = \frac{1}{-1+2} = 1. \quad (2.23)$$

For c_{11}, the formula is more complex. First, go back to $(s+1)^2X(s) = 1/(s+2)$, and take

$$c_{11} = \left[\frac{d}{ds}(s+1)^2X(s)\right]_{s=-1} = \left[\frac{d}{ds}\frac{1}{s+2}\right]_{s=-1} = \left[\frac{-1}{(s+2)^2}\right]_{s=-1} = -1. \quad (2.24)$$

Finally, the time-domain signal is given by

$$x(t) = c_{11}e^{-t} + c_{12}te^{-t} + c_{21}e^{-2t} = -e^{-t} + te^{-t} + e^{-2t}. \qquad (2.25)$$

(1-b) Example with real poles by clearing fractions

$$\begin{aligned} X(s) &= \frac{1}{(s+1)^2(s+2)} \\ &= \frac{c_{11}(s+1)(s+2) + c_{12}(s+2) + c_{21}(s+1)^2}{(s+1)^2(s+1)}. \end{aligned} \qquad (2.26)$$

Equating numerators, one finds

$$c_{11}s^2 + c_{11}3s + 2c_{11} + c_{12}s + 2c_{12} + c_{21}s^2 + 2c_{21}s + c_{21} = 1, \qquad (2.27)$$

which leads to the system of equations

$$\begin{aligned} s^2 \text{ term:} & \qquad 0 = c_{11} + c_{21} \\ s^1 \text{ term:} & \qquad 0 = 3c_{11} + c_{12} + 2c_{21} \\ s^0 \text{ term:} & \qquad 1 = 2c_{11} + 2c_{12} + c_{21} \end{aligned} \qquad (2.28)$$

or, in matrix form,

$$\begin{pmatrix} 1 & 0 & 1 \\ 3 & 1 & 2 \\ 2 & 2 & 1 \end{pmatrix} \begin{pmatrix} c_{11} \\ c_{12} \\ c_{21} \end{pmatrix} = \begin{pmatrix} 0 \\ 0 \\ 1 \end{pmatrix}. \qquad (2.29)$$

The system is a set of linear equations with real coefficients. It turns out that, when the partial fraction expansion is set-up correctly, the system always has as many equations as unknowns, and always has a unique solution. In this case, the solution is $c_{11} = -1$, $c_{12} = 1$, $c_{21} = 1$, as found earlier.

(2-a) Example with complex poles by residue method

$$\begin{aligned} X(s) &= \frac{2(s+1)}{s(s^2+2s+2)} \qquad \text{with poles at: } s = 0, \ s = -1 \pm j \\ &= \frac{c_{11}}{s} + \frac{c_{21}}{s+1-j} + \frac{c_{31}}{s+1+j} \qquad \text{with: } c_{31} = c_{21}^*. \quad (2.30) \end{aligned}$$

The coefficients are given by

$$
\begin{aligned}
c_{11} &= [sX(s)]_{s=0} = \left[\frac{2(s+1)}{s^2+2s+2}\right]_{s=0} = 1, \\
c_{21} &= [(s+1-j)X(s)]_{s=-1-j} = \left[\frac{2(s+1)}{s(s+1+j)}\right]_{s=-1+j} \\
&= \frac{2j}{(-1+j)(2j)} = \frac{1}{-1+j} = \frac{-1-j}{2}, \\
c_{31} &= \frac{-1+j}{2} \qquad (= c_{21}^*).
\end{aligned}
\tag{2.31}
$$

The time-domain function is

$$
\begin{aligned}
x(t) &= 1 + 2\operatorname{Re}(c_{21})e^{\operatorname{Re}(p_{21})t}\cos\left(\operatorname{Im}(p_{21})t\right) - 2\operatorname{Im}(c_{21})e^{\operatorname{Re}(p_{21})t}\sin\left(\operatorname{Im}(p_{21})t\right) \\
&= 1 - e^{-t}\cos t + e^{-t}\sin t.
\end{aligned}
\tag{2.32}
$$

Note that if the pole p_{31} is chosen instead of p_{21} in (2.32), the result remains the same.

(2-b) Example with complex poles by clearing fractions

$$
\begin{aligned}
X(s) &= \frac{2(s+1)}{s(s^2+2s+2)} \qquad \text{with poles at: } s=0,\ s=-1\pm j \\
&= \frac{c_{11}}{s} + \frac{c_{21}}{s+1-j} + \frac{c_{31}}{s+1+j} \qquad \text{with: } c_{31} = c_{21}^* \\
&= \frac{c_{11}(s+1-j)(s+1+j) + c_{21}s(s+1+j) + c_{31}s(s+1-j)}{s\left(s^2+2s+2\right)}.
\end{aligned}
\tag{2.33}
$$

Equating the numerators, one finds

$$
c_{11}s^2 + c_{11}2s + c_{11}2 + c_{21}s^2 + c_{21}(1+j)s + c_{31}s^2 + c_{31}s(1-j) = 2s+2
\tag{2.34}
$$

and the system of equations is

$$
\begin{pmatrix} 1 & 1 & 1 \\ 2 & (1+j) & (1-j) \\ 2 & 0 & 0 \end{pmatrix}
\begin{pmatrix} c_{11} \\ c_{21} \\ c_{31} \end{pmatrix}
= \begin{pmatrix} 0 \\ 2 \\ 2 \end{pmatrix}.
\tag{2.35}
$$

The result is a system of linear equations with complex coefficients, whose solution is identical to the one found in (2-a).

Complex arithmetic can be avoided by defining as unknowns the real parts and imaginary parts of the variables when the residue is complex. Specifically, let

$$
c_{21} = f_{21} + jg_{21}.
\tag{2.36}
$$

Then

$$X(s) = \frac{c_{11}}{s} + \frac{c_{21}}{s+1-j} + \frac{c_{21}^*}{s+1+j}$$

$$= \frac{c_{11}}{s} + \frac{(c_{21} + c_{21}^*)(s+1) + j(c_{21} - c_{21}^*)}{s^2 + 2s + 2}$$

$$= \frac{c_{11}(s^2 + 2s + 2) + (2f_{21}(s+1) - 2g_{21})s}{s(s^2 + 2s + 2)}, \tag{2.37}$$

which gives

$$c_{11}s^2 + 2sc_{11} + 2c_{11} + f_{21}2s^2 + f_{21}2s - 2g_{21}s = 2s + 2 \tag{2.38}$$

and the system of equations

$$\begin{pmatrix} 1 & 2 & 0 \\ 2 & 2 & -2 \\ 2 & 0 & 0 \end{pmatrix} \begin{pmatrix} c_{11} \\ f_{21} \\ g_{21} \end{pmatrix} = \begin{pmatrix} 0 \\ 2 \\ 2 \end{pmatrix}. \tag{2.39}$$

Solving the system produces $f_{21} = -1/2$, $g_{21} = -1/2$, $c_{11} = 1$, which (again) yields the same result as (2-a).

(3-a) Example with repeated complex poles by residue method

$$X(s) = \frac{1}{(s^2 + 2s + 2)^2} \qquad \text{with double poles at } s = -1 \pm j$$

$$= \frac{c_{11}}{s+1-j} + \frac{c_{11}^*}{s+1+j} + \frac{c_{12}}{(s+1-j)^2} + \frac{c_{12}^*}{(s+1+j)^2}. \tag{2.40}$$

Using the formula gives

$$c_{12} = \left[\frac{(s+1-j)^2}{(s^2 + 2s + 2)^2} \right]_{s=-1+j} = \left[\frac{1}{(s+1+j)^2} \right]_{s=-1+j}$$

$$= \frac{1}{(2j)^2} = -\frac{1}{4},$$

$$c_{11} = \left[\frac{d}{ds} \frac{1}{(s+1+j)^2} \right]_{s=-1+j} = \left[\frac{-2}{(s+1+j)^3} \right]_{s=-1+j}$$

$$= \frac{-2}{(2j)^3} = -\frac{j}{4}, \tag{2.41}$$

so that the time function is

$$x(t) = 2\operatorname{Re}(c_{11}) e^{-t} \cos(t) - 2\operatorname{Im}(c_{11}) e^{-t} \sin(t)$$

$$+ 2\operatorname{Re}(c_{12}) te^{-t} \cos(t) - 2\operatorname{Im}(c_{12}) te^{-t} \sin(t)$$

$$= \frac{1}{2} e^{-t} \sin(t) - \frac{1}{2} te^{-t} \cos(t). \tag{2.42}$$

(3-b) Example with repeated complex poles by clearing fractions
The example starts as before, but the complex conjugate terms are grouped together as follows

$$X(s) = \frac{1}{\left(s^2 + 2s + 2\right)^2} \qquad \text{with double poles at } p_1 = -1 + j \text{ and } p_2 = -1 - j$$

$$= \frac{c_{11}}{s + 1 - j} + \frac{c_{11}^*}{s + 1 + j} + \frac{c_{12}}{(s + 1 - j)^2} + \frac{c_{12}^*}{(s + 1 + j)^2}$$

$$= \frac{c_{11}}{s + 1 - j} + \frac{c_{11}^*}{s + 1 + j} + \frac{c_{12}}{s^2 + 2s - 2j(s + 1)} + \frac{c_{12}^*}{s^2 + 2s + 2j(s + 1)}.$$

$$\tag{2.43}$$

Defining

$$c_{11} = f_{11} + jg_{11}, \quad c_{12} = f_{12} + jg_{12}, \tag{2.44}$$

the transform becomes

$$X(s) = \frac{2f_{11}(s + 1) - 2g_{11}}{s^2 + 2s + 2} + \frac{2f_{12}(s^2 + 2s) - 4g_{12}(s + 1)}{(s^2 + 2s + 2)^2}. \tag{2.45}$$

Equating the numerators for $X(s)$ gives

$$\left(2f_{11}(s + 1) - 2g_{11}\right)\left(s^2 + 2s + 2\right) + 2f_{12}(s^2 + 2s) - 4g_{12}(s + 1) = 1, \tag{2.46}$$

and

$$s^3\left(2f_{11}\right) + s^2\left(2f_{11} + 4f_{11} - 2g_{11} + 2f_{12}\right)$$
$$+ s\ \left(4f_{11} + 4f_{11} - 4g_{11} + 4f_{12} - 4g_{12}\right)$$
$$+ \left(4f_{11} - 4g_{11} - 4g_{12}\right) = 1, \tag{2.47}$$

which corresponds to the system of equations

$$\begin{pmatrix} 2 & 0 & 0 & 0 \\ 6 & -2 & 2 & 0 \\ 8 & -4 & 4 & -4 \\ 4 & -4 & 0 & -4 \end{pmatrix} \begin{pmatrix} f_{11} \\ g_{11} \\ f_{12} \\ g_{12} \end{pmatrix} = \begin{pmatrix} 0 \\ 0 \\ 0 \\ 1 \end{pmatrix}. \tag{2.48}$$

The solution is $f_{11} = 0$, $f_{12} = -1/4$, $g_{11} = -1/4$, $g_{12} = 0$, which gives the same result as (3-a).

2.2.5 Non-strictly-proper transforms

The partial fraction expansion technique assumed that $X(s) = N(s)/D(s)$, where $N(s)$ and $D(s)$ are polynomials in s and $\deg N(s) < \deg D(s)$. By definition:

- a function that is a ratio of polynomials is called a *rational* function of s.

- a rational function of s such that $\deg N(s) < \deg D(s)$ is called a *strictly proper* function of s.

- a rational function of s such that $\deg N(s) \leqslant \deg D(s)$ is called a *proper* function of s.

In some cases, one may encounter a transform $X(s) = N(s)/D(s)$, with $\deg N(s) \geqslant \deg D(s)$. Such transform can be inverted using partial fraction expansions, with a preliminary step. First, using polynomial division, one finds $Q(s)$, $R(s)$ such that $N(s) = D(s)Q(s) + R(s)$ and $\deg R(s) < \deg D(s)$ ($Q(s)$ and $R(s)$ are the quotient and the remainder of the division of $N(s)$ by $D(s)$, respectively). As a result

$$X(s) = Q(s) + \frac{R(s)}{D(s)}. \tag{2.49}$$

The second term is a strictly proper rational function of s, which can be inverted using the partial fraction expansion procedure described earlier. The first term is a polynomial $Q(s) = q_0 + q_1 s + q_2 s^2 + ...$, whose associated time function is

$$q(t) = q_0\, \delta(t) + q_1 \frac{d}{dt}\left(\delta(t)\right) + q_2 \frac{d^2}{dt^2}\left(\delta(t)\right) + ... \tag{2.50}$$

In other words, $q(t)$ is a linear combination of the delta function and its derivatives. While the case of non-strictly-proper rational functions of s may be addressed easily in this manner, it is rarely meaningful in practice.

2.3 Properties of signals

2.3.1 Existence of terms in the partial fraction expansion

Important properties of the time function associated with a rational transform $X(s)$ may be derived from the knowledge of the procedure of partial fraction expansion. The expansion does not need to be performed: only the locations of the poles are needed. Consider the example

$$X(s) = \frac{N(s)}{(s+1)^2(s+2)}, \tag{2.51}$$

where $N(s)$ is an arbitrary polynomial. A partial fraction expansion will produce

$$X(s) = \frac{c_{11}}{s+1} + \frac{c_{12}}{(s+1)^2} + \frac{c_{21}}{s+2}. \tag{2.52}$$

Without computing the coefficients, one can predict that $x(t)$ will be a linear combination of the functions e^{-t}, te^{-t}, and e^{-2t}. The result is independent of $N(s)$. Further, if $N(s)$ does not have a root that is identical to a denominator root (pole/zero cancellation), the functions te^{-t}, and e^{-2t} must be present in the expansion. Indeed, the coefficients c_{12} and c_{21} cannot be zero because

$$
\begin{aligned}
c_{12} &= \left[(s+1)^2 X(s)\right]_{s=-1} = 0 \Leftrightarrow [N(s)]_{s=-1} = 0, \\
c_{21} &= \left[(s+2)X(s)\right]_{s=-2} = 0 \Leftrightarrow [N(s)]_{s=-2} = 0,
\end{aligned}
\tag{2.53}
$$

and there would have to be a pole/zero cancellation in $X(s)$, *i.e.*, a common root to $N(s)$ and $D(s)$.

In general, if there is no common root between $N(s)$ and $D(s)$, the residue formula implies that the coefficient associated to the highest power of a given pole must always be nonzero. Lower order terms may not be present. A similar conclusion may be drawn regarding the corresponding time function. The property is made precise in the following fact.

Fact - Nature of the time function associated with a rational X(s)
Consider a signal $x(t)$ with a transform $X(s) = N(s)/D(s)$, where $N(s)$ and $D(s)$ are polynomials in s with real coefficients having no common roots and such that $\deg N(s) < \deg D(s)$. Let the roots of $D(s) = 0$ be of the form $p = a \pm jb$, with multiplicity r. Then, the signal $x(t)$ is the linear combination of a set of time functions corresponding to the poles. For a given pole, the functions are $e^{at}\cos(bt+\phi)$, $te^{at}\cos(bt+\phi)$,..., and $t^{r-1}e^{at}\cos(bt+\phi)$, where ϕ is some phase angle. If the pole is real, the functions are e^{at}, te^{at}, ..., $t^{r-1}e^{at}$. The terms with the highest power of t (that is, $t^{r-1}e^{at}$ or $t^{r-1}e^{at}\cos(bt+\phi)$) **must be present** (*i.e.*, must have nonzero coefficients).
Example: consider the signal with transform

$$X(s) = \frac{s+1}{s^4 \left(s^2 + 4s + 13\right)^2 (s-10)}, \tag{2.54}$$

with poles at $s = 0$ (repeated 4 times), $s = -2 \pm 3j$ (repeated twice), and $s = 10$. Without performing a partial fraction expansion, we can say that $x(t)$ is a linear combination of the functions: 1, t, t^2, t^3, $e^{-2t}\cos(3t + \phi)$, $te^{-2t}\cos(3t + \phi)$, and e^{10t}. The coefficients of t^3, $te^{-2t}\cos(3t + \phi)$, and e^{10t} must be nonzero.

2.3.2 Boundedness and convergence of signals

The results derived from the partial fraction expansion procedure can be further developed to obtain statements about the boundedness and convergence of a signal with a rational transform. In particular, the results allow one to determine whether the final value theorem can be used, based on an inspection of the poles. We begin by recalling the definition of a bounded signal.

Definition - Bounded signal: a signal $x(t)$ is said to be *bounded* if there exists $M \geqslant 0$ such that

$$|x(t)| \leqslant M \qquad \text{for all } t \geqslant 0 \tag{2.55}$$

Conversely, a signal is said to be *unbounded* if no such bound can be found (meaning that, for all M, there exists some t such that $|x(t)| > M$).

The following definitions will also be useful

$$s \in \text{ open left half-plane (OLHP) } \Leftrightarrow \operatorname{Re}(s) < 0$$
$$s \in \text{ open right half-plane (ORHP) } \Leftrightarrow \operatorname{Re}(s) > 0$$
$$s \in \text{ closed left half-plane (CLHP) } \Leftrightarrow \operatorname{Re}(s) \leqslant 0$$
$$s \in \text{ closed right half-plane (CRHP) } \Leftrightarrow \operatorname{Re}(s) \geqslant 0$$
$$s \in j\omega - \text{axis} \Leftrightarrow \operatorname{Re}(s) = 0$$

Fact - Conditions for boundedness and convergence of signals with rational transforms

Assume that the signal $x(t)$ has a rational transform $X(s) = N(s)/D(s)$, where $N(s)$ and $D(s)$ are polynomials in s with real coefficients having no common roots and such that $\deg N(s) < \deg D(s)$. Then:

(a) The signal $x(t)$ is bounded if and only if the roots of $D(s)$ are either in the open left half-plane or are non-repeated roots on the $j\omega-$axis.

(b) The signal $x(t)$ has a limit for $t \to \infty$ if and only if the roots of $D(s)$ are in the open left half-plane, with the possible exception of a single root at the origin $(s = 0)$. If the limit exists, $\lim_{t\to\infty} x(t) = \lim_{s\to 0} sX(s)$.

(c) The signal $x(t)$ converges to zero as $t \to \infty$ if and only if all the roots of $D(s)$ are in the open left half-plane.

Proof: the proof is based on the properties of the individual terms of the partial fraction expansion. A term $t^{r-1}e^{at}\cos(bt+\phi)$ has the properties that:

(a) it is bounded if and only if $a < 0$ or $a = 0$ and $r = 1$ (the pole is in the OLHP or is non-repeated on the $j\omega$-axis).

(b) it has a limit if and only if $a < 0$ or $a = 0$, $b = 0$, and $r = 1$ (the pole is in the OLHP or is non-repeated at $s = 0$).

(c) it converges to zero if and only if $a < 0$ (the pole is in the OLHP).

The other elements of the proof are that the functions with the highest powers of t must be present in the expansion and that a function in the expansion cannot be cancelled by a combination of other functions (the functions are linearly independent).

Examples

1. $\dfrac{s-1}{(s+1)^2}$ bounded, converges to 0.

2. $\dfrac{s+1}{(s^2+4)}$ bounded, does not converge.

3. $\dfrac{s+1}{s^2}$ unbounded.

4. $\dfrac{s+5}{s(s+1)}$ bounded, converges to 5.

5. $\dfrac{s^2-1}{(s^2+16)^2}$ unbounded.

6. $\dfrac{1}{s^2-1}$ unbounded.

2.3.3   Non-strictly-proper transforms

The fact assumed that $X(s) = N(s)/D(s)$, where $N(s)$ and $D(s)$ are polynomials in s and $\deg N(s) < \deg D(s)$. For $\deg N(s) \geqslant \deg D(s)$, it was found earlier that

$$X(s) = Q(s) + X_{sp}(s), \tag{2.56}$$

where $X_{sp}(s)$ is a strictly proper function of s and $Q(s)$ is a polynomial in s. Further, $Q(s)$ corresponds to a time-domain function that is a linear combination of impulses and derivatives of impulses. Therefore, we have that

$$X(s) = \frac{N(s)}{D(s)} \quad \text{with } \deg N(s) \geqslant \deg D(s) \Rightarrow x(t) \text{ is unbounded.}$$
$$\tag{2.57}$$

2.4 Problems

Problem 2.1: Using both methods for partial fraction expansion (residue and clearing fractions), find the signals whose Laplace transforms are given by

(a) $X(s) = \dfrac{2s}{(s+2)(s-2)}$

(b) $X(s) = \dfrac{2s+1}{s^2(s+1)^2}$,

(c) $X(s) = \dfrac{s^3 + 2s^2 + 2s + 5}{s^2(s^2 + 2s + 5)}$

(d) $X(s) = \dfrac{4s - 8}{(s^2 + 4)^2}$.   (2.58)

Problem 2.2: (a) Consider a signal with Laplace transform

$$X(s) = \frac{s^2 + 4}{s^3(s^2 + 2s + 5)^2}.$$ (2.59)

Give the form of $x(t)$ that would result from a partial fraction expansion. Express the signal as a linear combination of time functions, but do not solve for the coefficients themselves. Indicate which of the coefficients may or may not turn out to be zero.

(b) repeat part (a) for

$$X(s) = \frac{s - 1}{(s+2)^4(s-3)^3(s+4)}.$$ (2.60)

Problem 2.3: For the signals whose Laplace transforms are given below, indicate whether the signals are bounded and, if so, whether $\lim\limits_{t\to\infty} x(t)$ exists. If the limit exists, give its value. Do not invert the Laplace transforms to obtain the results.

(a) $X(s) = \dfrac{10}{(s+1)^{10}}$

(b) $X(s) = \dfrac{(s-1)}{s(s+2)}$,

(c) $X(s) = \dfrac{1}{s^2(s+2)}$

(d) $X(s) = \dfrac{5}{s(s+1)^2}$,

(e) $X(s) = \dfrac{3}{s(s^2+4)}$

(f) $X(s) = \dfrac{3}{s(s^2+4)^2}$,

(g) $X(s) = \dfrac{2(s-1)}{(s^2+2s+1)(s+3)}$

(h) $X(s) = \dfrac{2(s-1)}{(s^2+2s+2)(s+3)}$,

(i) $X(s) = \dfrac{2(s-1)}{(s^2+2s+2)^2(s+3)}$.   (2.61)

Problem 2.4: (a) Is a signal $x(t)$ considered bounded if $x(t) < 2^{16}$?

(b) How fast does $x(t) = e^{-100t}\cos(10,000t)$ converge to zero?

(c) Is the signal $X(s) = \dfrac{(s-1)^2}{s(s^2+1)(s^2+4)}$ bounded? Does it converge to a steady-state value? If so, to what value?

Problem 2.5: (a) List the time functions that would originate from a partial fraction expansion of

$$X(s) = \frac{s+1}{(s^2+2s+5)^2(s^2+4)^3}.$$ (2.62)

Express the result in terms of real functions.

(b) Give the real time function $x(t)$ whose transform is

$$X(s) = \frac{1+2j}{(s+2+3j)^3} + \frac{1-2j}{(s+2-3j)^3}.$$ (2.63)

Problem 2.6: (a) Give the response $y(t)$ of a system with transfer function

$$P(s) = \frac{1}{s^2(s+1)}$$ (2.64)

and with constant input $x(t) = 1$.

(b) An input $x(t) = \sin(2t)$ is applied to a system with transfer function

$$P(s) = \frac{s-1}{s(s^2+4s+8)^3(s^2+4)}.$$ (2.65)

List all the functions that appear in the partial fraction expansion of $Y(s)$, without calculating the coefficients of the expansion. Express the result in terms of real functions, and indicate which functions must have nonzero coefficients.

Problem 2.7: (a) List the real time functions that would originate from a partial fraction expansion of the response of the system

$$P(s) = \frac{s+1}{s(s+100+33j)^2(s+100-33j)^2}$$ (2.66)

to a step input. Indicate which functions may and may not have zero coefficients in the partial fraction expansion.

(b) Repeat part (a) for an input $x(t) = \cos(33t)$.

Chapter 3

Continuous-time systems

3.1 Transfer functions and interconnected systems

3.1.1 Transfer function of a system

A *continuous-time system* is an operator that transform a continuous-time signal into another continuous-time signal. For example, consider the simple circuit shown in Fig. 3.1. The circuit is described by

$$v(t) = L\frac{di(t)}{dt} + Ri(t), \tag{3.1}$$

which may be transformed to the s-domain by using the properties of the Laplace transform

$$V(s) = sLI(s) - Li(0) + RI(s) \Rightarrow I(s) = \frac{1}{sL + R}V(s) + \frac{Li(0)}{sL + R}. \tag{3.2}$$

Consider $V(s)$ to be the input to the system, and $I(s)$ to be the output. The first term in the expression for $I(s)$ is due to the input, while the second term

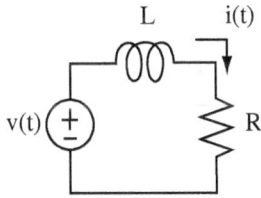

Figure 3.1: RL circuit for example

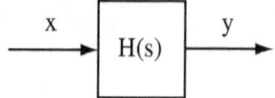

Figure 3.2: Linear time-invariant system

is due to the initial current in the inductor. This initial current is considered to be an *initial condition* of the system, or *initial state*. For the time being, we will let $i(0) = 0$. Then

$$\frac{I(s)}{V(s)} = \frac{1}{sL + R} = \frac{1/L}{s + R/L}. \tag{3.3}$$

By definition

$$H(s) = \frac{I(s)}{V(s)}, \tag{3.4}$$

where $H(s)$ is called the *transfer function* of the system. Note that

$$H(s) = \frac{1/L}{s + R/L} = \text{Laplace transform} \left(\frac{1}{L} e^{-(R/L)t} \right)$$

$$\triangleq \text{Laplace transform } (h(t)), \tag{3.5}$$

where $h(t)$ is the *impulse response* of the system. Indeed for $v(t) = \delta(t)$, $V(s) = 1$, $I(s) = H(s)$, and $i(t) = h(t)$.

In general, a linear time-invariant system with input $x(t)$ and output $y(t)$ may be represented by the block diagram of Fig. 3.2. This representation means that

$$Y(s) = H(s)X(s). \tag{3.6}$$

A system is completely described by $H(s)$, since the response to any input signal can be computed from the knowledge of the signal and of the transfer function of the system. When

$$H(s) = \frac{N(s)}{D(s)}, \tag{3.7}$$

where $N(s)$ and $D(s)$ are polynomials, the roots of $N(s)$ are called the *zeros* of the system, and the roots of $D(s)$ are called the *poles* of the system.

In the time domain

$$y(t) = h(t) * x(t), \tag{3.8}$$

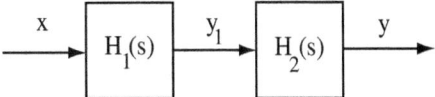

Figure 3.3: Cascade system

where "∗" denotes the *convolution* operation [26], or

$$h(t) * x(t) = \int_{-\infty}^{\infty} h(\tau)x(t-\tau)d\tau. \tag{3.9}$$

Knowledge of $h(t)$ is equivalent to knowledge of $H(s)$. The advantage of $H(s)$ is that the rather complicated time-domain convolution operation is replaced by a simple multiplication in the s-domain. Note that, with the unilateral definition of the Laplace transform (integration from $t = 0$ to ∞), the equivalence between convolution in the time domain and multiplication in the s-domain is based on the assumption that $h(t)$ and $x(t)$ are zero for $t < 0$.

3.1.2 Cascade systems

The replacement of convolution by a simple multiplication in the s-domain allows one to rapidly compute the transfer functions of interconnected systems. A cascade system is shown in Fig. 3.3. The overall transfer function is obtained as follows

$$\begin{aligned} Y_1(s) &= H_1(s)X(s) \\ Y(s) &= H_2(s)Y_1(s), \end{aligned} \tag{3.10}$$

so that

$$\begin{aligned} Y(s) &= H_2(s)H_1(s)X(s) \\ &= H(s)X(s) \quad \text{for} \quad H(s) = H_2(s)H_1(s). \end{aligned} \tag{3.11}$$

In other words, the transfer function of a cascade system is the product of the two transfer functions. If we let

$$H_1(s) = \frac{N_1(s)}{D_1(s)} \qquad H_2(s) = \frac{N_2(s)}{D_2(s)}, \tag{3.12}$$

the result is

$$H(s) = \frac{N_1(s)N_2(s)}{D_1(s)D_2(s)}. \tag{3.13}$$

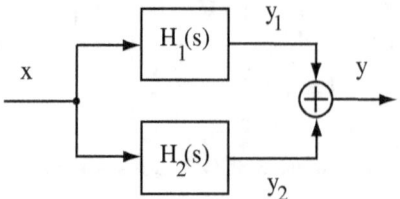

Figure 3.4: Parallel system

Therefore, unless there are pole/zero cancellations, the poles of $H(s)$ are the union of the poles of $H_1(s)$ and $H_2(s)$, and the zeros of $H(s)$ are the union of the zeros of $H_1(s)$ and $H_2(s)$.

3.1.3 Parallel systems

A parallel system is shown in Fig. 3.4. The output of the system is given by

$$
\begin{aligned}
Y(s) &= Y_1(s) + Y_2(s) = H_1(s)X(s) + H_2(s)X(s) \\
&= (H_1(s) + H_2(s))\, X(s), \tag{3.14}
\end{aligned}
$$

so that the overall transfer function is the sum of the two transfer functions

$$H(s) = H_1(s) + H_2(s). \tag{3.15}$$

With

$$H_1(s) = \frac{N_1(s)}{D_1(s)} \qquad H_2(s) = \frac{N_2(s)}{D_2(s)}, \tag{3.16}$$

one has

$$H(s) = \frac{N_1(s)D_2(s) + D_1(s)N_2(s)}{D_1(s)D_2(s)}. \tag{3.17}$$

The poles of $H(s)$ are (again) the union of the poles of $H_1(s)$ and $H_2(s)$, but the zeros are the roots of $N_1(s)D_2(s) + D_1(s)N_2(s) = 0$ (except for possible cancellations). A zero that is common to both $H_1(s)$ and $H_2(s)$ is also a zero of $H(s)$.

3.1.4 Feedback system

A feedback system is shown in Fig. 3.5. The output of the system is given by

$$
\begin{aligned}
Y(s) &= H_1(s)E(s) \\
&= H_1(s)X(s) - H_1(s)H_2(s)Y(s), \tag{3.18}
\end{aligned}
$$

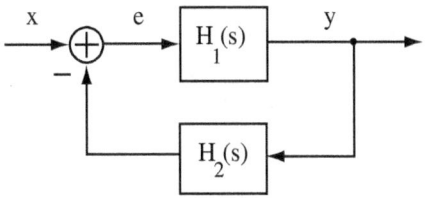

Figure 3.5: Feedback system

which implies that

$$(1 + H_1(s)H_2(s)) Y(s) = H_1(s)X(s). \tag{3.19}$$

The result is the overall transfer function

$$H(s) = \frac{H_1(s)}{1 + H_1(s)H_2(s)}. \tag{3.20}$$

This is an important formula for the design of feedback systems. The overall transfer function is the ratio of the so-called *forward path transfer function* $H_1(s)$ and the number 1 plus the so-called *loop transfer function* $H_1(s)H_2(s)$. $H_2(s)$ is called the *feedback path transfer function*.

In terms of the original poles and zeros

$$H(s) = \frac{N_1(s)D_2(s)}{N_1(s)N_2(s) + D_1(s)D_2(s)}. \tag{3.21}$$

The poles of the overall system are the roots of $N_1(s)N_2(s) + D_1(s)D_2(s) = 0$ and are called the *closed-loop poles*. In general, these poles are different from the *open loop poles* (the poles of $H_1(s)$ and $H_2(s)$). Indeed, if p is an open-loop pole, $D_1(p) = 0$ or $D_2(p) = 0$. To be a closed-loop pole, it would need to satisfy $N_1(p) = 0$ or $N_2(p) = 0$. Therefore, unless there is a pole/zero cancellation in the product $H_1(s)H_2(s)$, the closed-loop poles are distinct from the open-loop poles. Finally, note that the zeros of $H(s)$ are the union of the zeros of the forward path and of the poles of the feedback path.

3.1.5   Block reduction method

The formulas for cascade, parallel, and feedback systems can be combined to find the transfer function of many interconnected systems. For example, consider the system of Fig. 3.6. The parallel blocks $H_1(s)$ and $H_2(s)$ can be replaced by $H_1(s) + H_2(s)$. In cascade with $H_4(s)$, the combined transfer function of the forward path is $(H_1(s) + H_2(s)) H_4(s)$. Then, using the feedback system formula, the overall transfer function is found to be

$$H(s) = \frac{(H_1(s) + H_2(s)) H_4(s)}{1 + (H_1(s) + H_2(s)) H_3(s) H_4(s)}. \qquad (3.22)$$

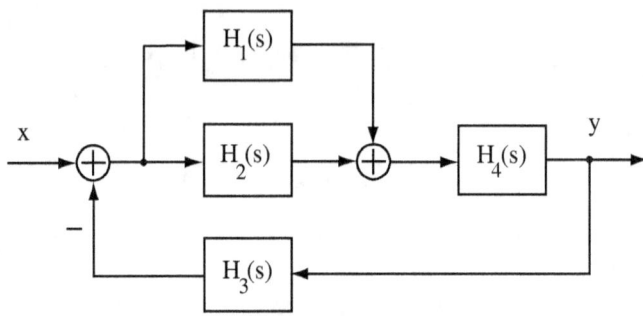

Figure 3.6: Interconnected system

If summation junctions are moved through equivalent transformations, even more problems can be solved. For example, consider the system of Fig. 3.7. The structure of the system is similar to Fig. 3.6, but none of the formulas can be applied directly. However, the summation junction between $H_2(s)$ and $H_4(s)$ can be moved to the output of $H_4(s)$ by replacing $H_1(s)$ by $H_1(s)H_4(s)$, as shown in Fig. 3.8. Then, the formula for cascade systems makes it possible to replace the two blocks $H_2(s)$ and $H_4(s)$ by a single block $H_2(s)H_4(s)$.

To proceed further, the output of $H_1(s)H_4(s)$ is moved past the feedback from y, requiring that one add a feedforward term from x as shown in Fig. 3.9. Then, the transfer function of the system can be computed to be

$$
\begin{aligned}
H(s) &= H_1(s)H_4(s) + (1 - H_1(s)H_3(s)H_4(s)) \frac{H_2(s)H_4(s)}{1 + H_2(s)H_3(s)H_4(s)} \\
&= \frac{N(s)}{1 + H_2(s)H_3(s)H_4(s)}, \qquad (3.23)
\end{aligned}
$$

where

$$
\begin{aligned}
N(s) &= H_1(s)H_4(s) + H_1(s)H_2(s)H_3(s)H_4(s)^2 + H_2(s)H_4(s) \\
&\quad - H_1(s)H_2(s)H_3(s)H_4(s)^2 \\
&= (H_1(s) + H_2(s)) H_4(s). \qquad (3.24)
\end{aligned}
$$

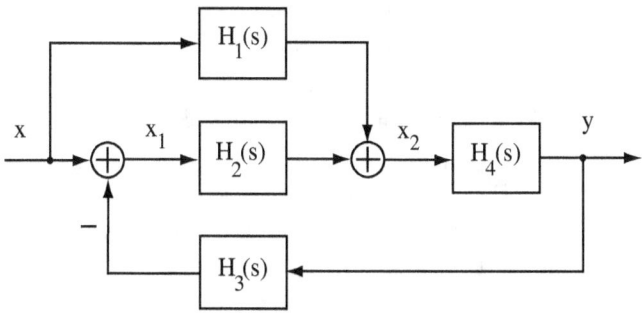

Figure 3.7: Interconnected system

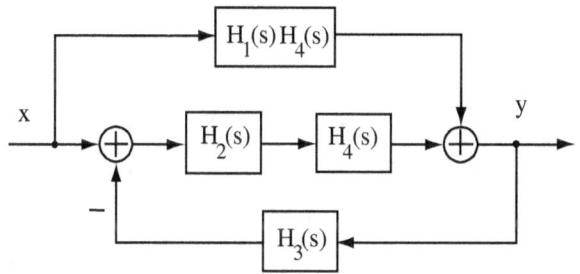

Figure 3.8: Block diagram after the first step

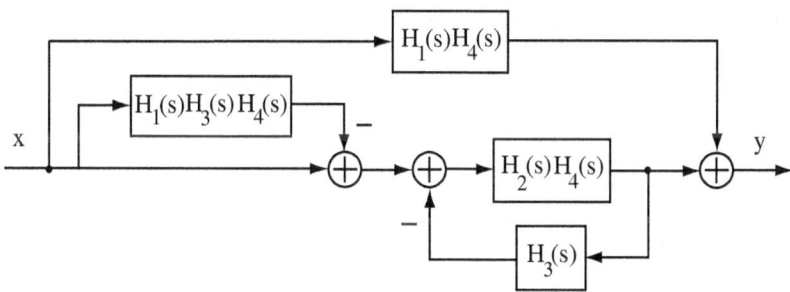

Figure 3.9: Block diagram after the second step

As seen in this example, the transfer function of many systems can be found by the block reduction method using the formulas for cascade, parallel, and feedback systems, and carefully transforming the diagrams into equivalent systems. In difficult cases, skillful manipulations of the diagram may be required. The so-called *Mason's rule* is a general procedure that can also be used and is often found in textbooks. However, it also requires some skill to apply. The next section presents a procedure that can be applied systematically and with less risk of error in complicated cases.

3.1.6   General interconnected systems

Interconnected systems are assumed to be composed of linear time-invariant systems and summing junctions, with a single input and a single output for which the transfer function must be obtained.

Procedure

1. Define n variables $X_i(s)$ after every summing junction.

2. Write equations for $Y(s)$ and for the $X_i(s)'s$, by reading the diagram. This step produces $n+1$ equations relating $X_i(s)$, $Y(s)$ and $X(s)$.

3. Eliminate the $X_i(s)'s$ until a single equation is left that relates $Y(s)$ and $X(s)$. This step is equivalent to solving $n+1$ linear equations in the $n+1$ unknowns $Y(s)$ and $X_i(s)$. The solution for $Y(s)$ gives the transfer function.

Example: consider the block diagram of Fig. 3.7. The equations are

$$
\begin{aligned}
X_1(s) &= X(s) - H_3(s)Y(s) \\
X_2(s) &= H_1(s)X(s) + H_2(s)X_1(s) \\
Y(s) &= H_4(s)X_2(s).
\end{aligned}
\tag{3.25}
$$

Eliminating the extra variables

$$
\begin{aligned}
Y(s) &= H_4(s)H_1(s)X(s) + H_4(s)H_2(s)X_1(s) \\
&= H_4(s)H_1(s)X(s) + H_4(s)H_2(s)X(s) \\
&\quad - H_4(s)H_2(s)H_3(s)Y(s),
\end{aligned}
\tag{3.26}
$$

and

$$(1 + H_4(s)H_2(s)H_3(s)) Y(s) = H_4(s) (H_1(s) + H_2(s)) X(s).$$
(3.27)

Therefore, the transfer function is

$$H(s) = \frac{H_4(s) (H_1(s) + H_2(s))}{1 + H_2(s)H_3(s)H_4(s)}.$$
(3.28)

The result is the same as was obtained through block reduction.

3.2 Stability

3.2.1 Definitions

In the previous chapter, we discussed the *boundedness* and *convergence* properties of *signals*. Now, we discuss the *stability* properties of *systems*. The two sets of properties are very much related, and rely heavily on concepts derived from partial fraction expansions. However, they concern distinct objects, namely *signals* and *systems*. We begin with a standard definition of stability and with a test for stability.

Definition - BIBO stability: a linear time-invariant system is called *bounded-input bounded-output stable* (*BIBO stable*) if the output is bounded for any bounded input. A system is called *BIBO unstable* if it is not BIBO stable.

Comments

(a) A system is *unstable* if there exists a bounded input signal such that the output is unbounded. The output of the system does not need to be unbounded for *all* bounded input signals: only one signal is sufficient. Further, the output of an unstable system may even be bounded for some unbounded input signals.

(b) The fact below states that a system is BIBO stable if all its poles are in the open left half-plane. Then, two types of unstable systems may be encountered: systems that have some repeated poles on the $j\omega-$axis and/or some poles in the open right half-plane, and systems that only have non-repeated poles on the $j\omega-$axis. In the first case, virtually every bounded input yields an unbounded output. In the second case, only well-chosen inputs produce an unbounded output.

3.2.2 Properties

Fact - BIBO stability of systems with rational transfer functions

A linear time-invariant system with $H(s) = N(s)/D(s)$ and $\deg N(s) \leqslant \deg D(s)$

is BIBO stable if and only if all the poles of the transfer function are in the open left half-plane $(\text{Re}(s) < 0)$.

Example 1: $H(s) = \dfrac{1}{(s+1)^2}$ is stable. The output is bounded for all bounded inputs.

Example 2: $H(s) = \dfrac{1}{s-1}$ is unstable. For $x(t) = u(t)$ (step input), the output is $y(t) = -1 + e^t$ and is unbounded. The output for $X(s) = (s-1)/(s+1)^2$ is bounded. However, all bounded inputs whose transforms do not have a zero at $s = 1$ will yield unbounded outputs.

Example 3: $H(s) = \dfrac{1}{(s^2+1)}$ is unstable. However, the only bounded inputs that yield an unbounded output are those which have poles at $s = \pm j$. If, for example, $x(t) = u(t)$, then $Y(s) = 1/(s(s^2+1)) = 1/s - s/(s^2+1)$, and $y(t) = 1 - \cos(t)$, which is bounded. On the other hand, if $x(t) = 2\cos(t)$, then $Y(s) = 2s/(s^2+1)^2$, and $y(t) = t\sin(t)$, which is unbounded.

Proof of the fact

The proof given below assumes that the input signals under consideration have rational transforms. However, the result is true in general.

(a) Pole condition $\Rightarrow$ BIBO stability. Recall that a signal is bounded if and only if the poles of its transform are in the OLHP or are non-repeated poles on the $j\omega$-axis. Since the poles of $Y(s)$ are the union of those of $X(s)$ and $H(s)$, and since all the poles of $H(s)$ are in the OLHP, $Y(s)$ will satisfy the condition for boundedness if $X(s)$ does.

(b) BIBO stability $\Rightarrow$ pole condition. The result is proved by contradiction: if the pole condition is not satisfied, there is a bounded input for which the output is unbounded. Many input signals may exist, but only one signal needs to be found. Two cases are considered. If $H(s)$ has some repeated poles on the $j\omega$-axis and/or some poles in the ORHP, let $x(t) = \cos(\omega_0 t)$, where $s = j\omega_0$ is any value that is not a zero of $H(s)$. The resulting output is unbounded, since the partial fraction expansion of the output must include unbounded signals corresponding to the poles of $H(s)$. If $H(s)$ has non-repeated poles on the $j\omega$-axis, only well-chosen signals produce unbounded outputs. In particular, a signal $x(t) = \cos(\omega_0 t)$ with a frequency that matches the location of one of the poles of $H(s)$ on the $j\omega$-axis yields an unbounded output, because the partial fraction expansion of the output must contain imaginary poles with multiplicity greater than 1.

3.2.3 Non-proper transfer functions

For a non-proper transfer function ($\deg N(s) > \deg D(s)$), the response to a step input includes impulses, so that

$$H(s) = \frac{N(s)}{D(s)} \quad \text{with } \deg N(s) > \deg D(s) \Longrightarrow H(s) \text{ is BIBO unstable.}$$
(3.29)

3.3 Responses to step inputs

3.3.1 General characteristics of step responses

Consider a system with transfer function $H(s) = N(s)/D(s)$. The *step response* of the system is the response to a step input

$$x(t) = x_m \, u(t),$$
(3.30)

where $u(t) = 0$ for $t < 0$ and $u(t) = 1$ for $t \geqslant 0$. In the s-domain, the step response is given by

$$Y(s) = H(s)\frac{x_m}{s} = \frac{x_m N(s)}{sD(s)}.$$
(3.31)

If the magnitude of the step is 1 ($x_m = 1$), the step response is $(1/s)H(s)$ and is called the unit step response. The unit step response is the integral of the impulse response. For this reason, the step response is often used to characterize dynamic systems. It contains complete information about the poles and zeros of the system and can be used to estimate the transfer function of the system. The step response is also important because, in many control applications, the control input changes abruptly and remains constant for long periods of time (set-point regulation).

General characteristics of the step response can be derived from knowledge of partial fraction expansions. *We assume that the system is BIBO stable.* Then, $H(s)$ cannot have a pole at the origin and the step response is a bounded signal. A partial fraction expansion of $Y(s)$ will lead to an expression of the form

$$Y(s) = \frac{H(0)x_m}{s} + \frac{N_1(s)}{D(s)},$$
(3.32)

where the first term is obtained from the residue formula and the second term groups all the contributions of the poles of $H(s)$ (with $N_1(s)$ some polynomial in s). $H(0)$ is the value of the transfer function evaluated at $s = 0$, and is finite by virtue of the stability assumption. Define

$$Y_{ss}(s) = \frac{H(0)x_m}{s} \qquad Y_{tr}(s) = \frac{N_1(s)}{D(s)},$$
(3.33)

where $Y_{ss}(s)$ is called the *steady-state response* of the system and $Y_{tr}(s)$ is called the *transient response* of the system. Because of the stability assumption on $H(s)$, the transient response is an exponentially decaying function in the time domain. The steady-state response is a constant signal

$$y_{ss}(t) = H(0)x_m. \tag{3.34}$$

$H(0)$ is called the *DC gain* or *steady-state gain* of the system. Assuming a constant input signal and neglecting the transient response terms, the DC gain is the ratio of the magnitude of the output signal to the magnitude of the input signal.

3.3.2   Examples

Example 1: consider a first-order system

$$H(s) = \frac{k}{s+a}. \tag{3.35}$$

The DC gain is k/a and, through a partial fraction expansion, the steady-state and transient responses can be found to be

$$y_{ss}(t) = \frac{k}{a}x_m \qquad y_{tr}(t) = -\frac{k}{a}x_m e^{-at}. \tag{3.36}$$

The constant $\tau = 1/a$ is usually referred to as the *time constant* of the system. $[dy/dt]_{t=0} = kx_m$, so that the intersection of the tangent to the response at $t = 0$ and the steady-state value $H(0)x_m = kx_m/a$ occurs for $t = 1/a = \tau$, the time constant of the system. These properties and the shape of the step response are shown in Fig. 3.10.   For $t = \tau$, $y(t) = H(0)x_m(1 - e^{-1})$, which implies that the output reaches 63% of its steady-state value after a time equal to the time constant. Often, a value of time equal to 4τ is taken to be the convergence time, or settling time. After that time, the output has reached 98% of its steady-state value.

Example 2: If the transfer function has a double real pole

$$H(s) = \frac{k}{(s+a)^2}, \tag{3.37}$$

a partial fraction expansion gives

$$y_{ss}(t) = \frac{k}{a^2}x_m \qquad y_{tr}(t) = -\frac{k}{a}x_m te^{-at} - \frac{k}{a^2}x_m e^{-at}. \tag{3.38}$$

The step responses of the first-order system with pole at $-a$ and of the second-order system with double pole at $s = -a$ are shown on Fig. 3.11. The responses

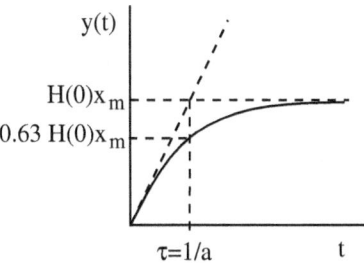

Figure 3.10: Step response of a first-order system

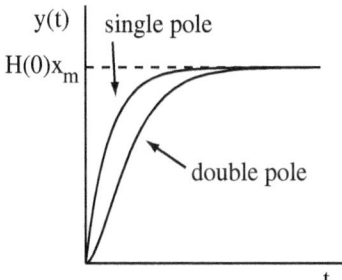

Figure 3.11: Step responses of a first-order system and of a second-order system with double pole at the same location

are qualitatively similar, but one can see that the delay of the response is roughly doubled. Also, the derivative at $t = 0$ is zero for the double pole. The slope is always zero if the number of poles exceeds the number of zeros by 2 or more.

Example 3: If the system has two distinct real poles

$$H(s) = \frac{k}{(s + a_1)(s + a_2)},$$
(3.39)

a partial fraction expansion gives

$$y_{ss}(t) = \frac{k}{a_1 a_2} x_m \qquad y_{tr}(t) = \frac{k}{a_1(a_1 - a_2)} x_m e^{-a_1 t} + \frac{k}{a_2(a_2 - a_1)} x_m e^{-a_2 t}.$$
(3.40)

The pole with the larger magnitude yields a term that converges faster to zero, so that the response is usually dominated by the contribution of the pole with smaller magnitude. Fig. 3.12 shows the response of a first-order system with pole

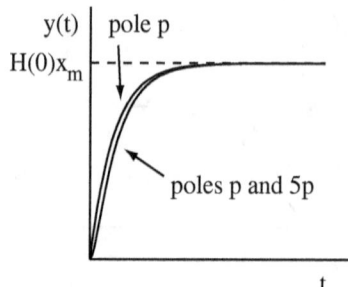

Figure 3.12: Step responses of a first-order system with pole p and of a second-order system with poles p and 5p

at $-a_1$ and of a second-order system with poles at $-a_1$ and $-a_2 = -5a_1$. Note that the responses are very close. The additional pole creates a small amount of delay, and the slope around $t = 0$ is zero for the second-order system. However, the first-order system is a good approximation of the second-order system, even though the additional pole is only 5 times greater.

Example 4: If the system has a pair of complex poles at $s = -a \pm jb$,

$$H(s) = \frac{k}{(s + a - jb)(s + a + jb)} = \frac{k}{s^2 + 2as + a^2 + b^2}, \qquad (3.41)$$

a partial fraction expansion gives

$$y_{ss}(t) = H(0)x_m \qquad y_{tr}(t) = -H(0)x_m e^{-at} \cos(bt) - H(0)x_m \frac{a}{b} e^{-at} \sin(bt), \qquad (3.42)$$

where

$$H(0) = \frac{k}{a^2 + b^2}. \qquad (3.43)$$

The step response typically exhibits oscillations associated with the sinusoidal components of the transient response. However, the magnitude of these oscillations depends on the rate of decay associated with the real part of the complex pole, as compared to the imaginary part of the pole. Fig. 3.13 shows the responses for several cases. If the imaginary part is larger than the real part of the pole in magnitude, oscillations are visible and produce a large overshoot in the response.

For a small ratio a/b, the response is approximately

$$y(t) \simeq H(0)x_m(1 - e^{-at} \cos(bt)), \qquad (3.44)$$

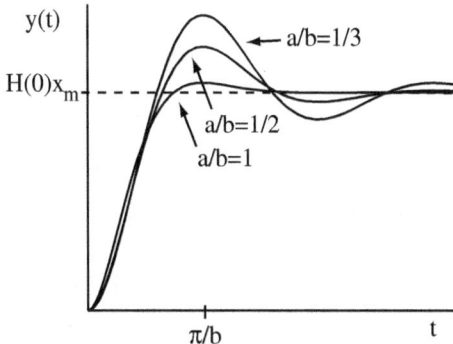

Figure 3.13: Step responses of systems with complex poles

so that the peak of the response occurs for $t \simeq \pi/b$. The percent overshoot is

$$PO\ (\%) \simeq 100e^{-a\pi/b}. \tag{3.45}$$

The formula gives 4% for $a/b = 1$, 20% for $a/b = 1/2$, and 35% for $a/b = 1/3$, which is consistent with the figure. It is typical to define the *damping factor* ζ and the *natural frequency* ω_n through

$$\zeta = \frac{a}{\sqrt{a^2 + b^2}}, \qquad \omega_n = \sqrt{a^2 + b^2}. \tag{3.46}$$

The natural frequency is the magnitude the pole, and the damping factor is the cos of the angle between the pole and the negative real axis in the s-domain (see Fig. 3.14).

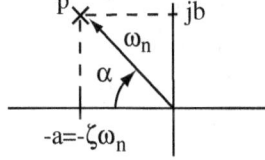

Figure 3.14: Definition of variables for complex poles

Systems with low damping are those for which

$$\frac{a}{b} < 1 \Leftrightarrow \zeta < 0.707. \tag{3.47}$$

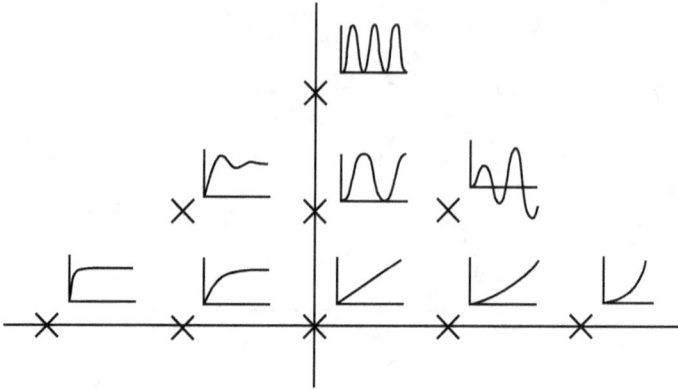

Figure 3.15: Step responses *vs.* pole locations

When the damping factor is small, $\zeta \simeq a/b$ and $\omega_n \simeq b$, so that the oscillation frequency is close to the natural frequency. An example of a system with a low damping is a slender robotic arm, such as the one used in the space shuttle.

3.3.3   Effect of poles and zeros on step responses

The characteristics of step responses for systems with single real and complex poles are shown in Fig. 3.15. Unstable systems are also considered, where the responses are unbounded. Although the responses of systems with more poles are not included, the components of the responses depend on the pole locations in a similar manner. Further, the responses are often characterized by the real pole or complex pole pair which is the farthest to the right of the s-plane, and is therefore called the *dominant pole*.

Sometimes, the zeros of the transfer function can have a significant effect on the step responses. This phenomenon occurs in particular when the magnitude of a zero is much smaller than the magnitude of the poles. Overshoot may occur even if the poles are real. Fig. 3.16 shows the response to be expected of a second-order system with two poles and one zero as the location of the zero is varied. Without the zero, no overshoot is observed. When the zero is at the origin, the DC gain of the system is zero. Therefore, the response of the system converges to zero in the steady-state and consists only of the transient response. If the zero is not at the origin, but is closer to the origin than the poles, the response converges to a nonzero value, but exhibits a significant overshoot (without oscillations). Fig. 3.16 also shows that, when the zero is in

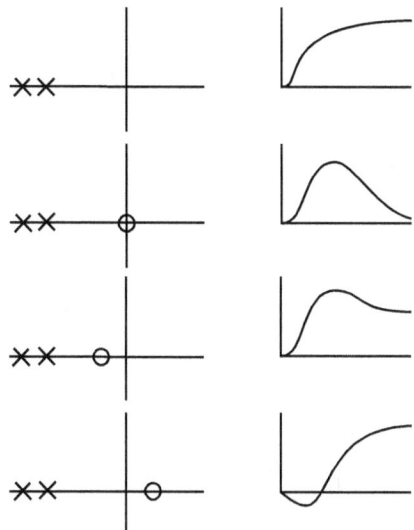

Figure 3.16: Effect of a zero on step responses

the right half-plane, the response may exhibit *undershoot*. Such characteristic is undesirable in control systems. Generally, zeros in the right half-plane are called *non-minimum-phase* zeros and are unfavorable for reasons including, but not limited to the transient characteristic discussed here. Zeros in the left half-plane are called *minimum-phase zeros*.

3.4   Responses to sinusoidal inputs

3.4.1   Definition and example

The *sinusoidal response* of a system is the response to an input signal

$$x(t) = x_m \cos(\omega_0 t + \alpha). \tag{3.48}$$

It is assumed that the system is BIBO stable, in order to guarantee the boundedness of the response. As an example, consider the RL circuit of Fig. 3.1. Let $R = 1\ \Omega$ and $L = 1$ H, and the input voltage $v(t) = \sin(t)$, so that

$$I(s) = \frac{1}{s+1}V(s) = \frac{1}{s+1}\frac{1}{s^2+1}. \tag{3.49}$$

A partial fraction expansion gives the time-domain function

$$i(t) = \frac{1}{2}e^{-t} - \frac{1}{2}\cos(t) + \frac{1}{2}\sin(t). \tag{3.50}$$

Similarly to step responses, the first term is an exponentially decaying term called the *transient response*. The last two terms are sinusoidal functions at the frequency of the input, and constitute the *steady-state response*. In other words

$$i(t) = i_{tr}(t) + i_{ss}(t), \tag{3.51}$$

with

$$i_{tr}(t) = \frac{1}{2}e^{-t}, \; i_{ss}(t) = -\frac{1}{2}\cos(t) + \frac{1}{2}\sin(t). \tag{3.52}$$

Alternatively, $i_{ss}(t)$ is also

$$i_{ss}(t) = \frac{1}{\sqrt{2}}\cos(t + 225°) = \frac{1}{\sqrt{2}}\sin(t - 45°), \tag{3.53}$$

which shows that the steady-state response is a sinusoid with the same frequency as the input signal, but different magnitude and phase. Note that it was assumed in the derivation of $I(s)$ that $i(0) = 0$ (the initial current in the inductor was zero). An interesting consequence of that assumption is that the signal $i(t)$ obtained here is the current that would appear in the RL circuit if the voltage source was connected at $t = 0$ (as if a switch was closed at $t = 0$). Although the input signal was taken to be a sinusoid, the value of the signal for $t < 0$ is not considered in the Laplace transform analysis and the initial condition $i(0)$ solely determines the state of the circuit at $t = 0$.

The transient, steady-state, and overall current waveforms are shown in Fig. 3.17 for the system $H(s) = 1/(s + 1)$ and an input $x(t) = \sin(3t)$. As expected, the current is zero at $t = 0$. An overcurrent of about 50% is observed within about a second. Such large transient currents are observed in power distribution systems, and need to be accounted for in the design of the components and their protection.

3.4.2   General characteristics of steady-state sinusoidal responses

We first consider the case where $x(t) = x_m \cos(\omega_0 t)$, so that the Laplace transform of the output is given by

$$Y(s) = H(s)x_m \frac{s}{s^2 + \omega_0^2}. \tag{3.54}$$

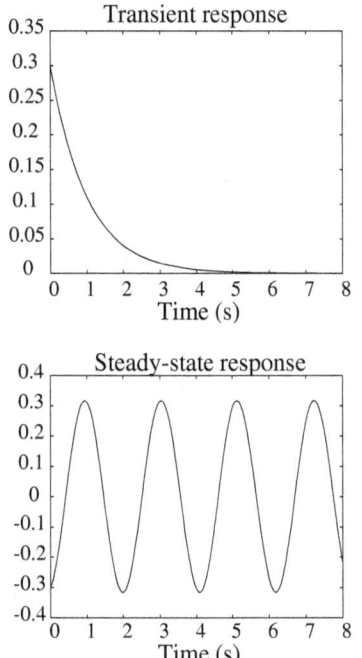

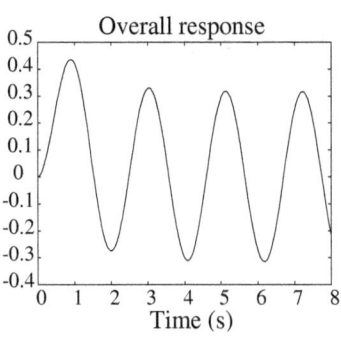

Figure 3.17: Transient, steady-state and overall current responses

A partial fraction expansion leads to terms of the form

$$Y(s) = \frac{1}{2}x_m \frac{H(j\omega_0)}{s - j\omega_0} + \frac{1}{2}x_m \frac{H(-j\omega_0)}{s + j\omega_0} + \frac{N_2(s)}{D(s)}, \qquad (3.55)$$

where the first two terms are obtained by the residue method and the last term groups all the components due to the poles of $H(s)$. $H(j\omega_0)$ is the value of the transfer function evaluated at $s = j\omega_0$.

As for the step response, assume that the system is BIBO stable and define

$$Y_{ss}(s) = \frac{1}{2}x_m \frac{H(j\omega_0)}{s - j\omega_0} + \frac{1}{2}x_m \frac{H(-j\omega_0)}{s + j\omega_0}, \qquad Y_{tr}(s) = \frac{N_2(s)}{D(s)}, \qquad (3.56)$$

where $Y_{ss}(s)$ and $Y_{tr}(s)$ are the steady-state and transient responses, respectively. In the time domain, the transient response is a signal that decays to zero exponentially because of the stability assumption on the system.

One has that $H(-j\omega_0) = H^*(j\omega_0)$ for a system with a real impulse response.

so that

$$Y_{ss}(s) = \frac{x_m}{s^2 + \omega_0^2} \left(s \frac{H(j\omega_0) + H^*(j\omega_0)}{2} + j\omega_0 \frac{H(j\omega_0) - H^*(j\omega_0)}{2} \right)$$

$$= \frac{x_m}{s^2 + \omega_0^2} \left(s \operatorname{Re}\left(H(j\omega_0)\right) - \omega_0 \operatorname{Im}\left(H(j\omega_0)\right) \right). \tag{3.57}$$

In the time domain,

$$y_{ss}(t) = \operatorname{Re}(H(j\omega_0))x_m \cos(\omega_0 t) - \operatorname{Im}(H(j\omega_0))x_m \sin(\omega_0 t),$$
$$\tag{3.58}$$

or

$$y_{ss}(t) = M x_m \cos(\omega_0 t + \phi), \tag{3.59}$$

where

$$M = |H(j\omega_0)|, \qquad \phi = \angle(H(j\omega_0)). \tag{3.60}$$

Equation (3.59) highlights the similarity between the input and output signals in the steady-state. $|H(j\omega_0)|$ is viewed as the gain of the system for sinusoidal inputs, because it is the ratio of the magnitude of the output signal to the magnitude of the input signal. The output signal is shifted in time with respect to the input signal, with the effect of a delay if $\phi < 0$ and of an advance if $\phi > 0$. Equation (3.60) shows that the gain and phase relating the input and output signals are given by the magnitude and angle of the complex number $H(j\omega_0)$. $H(j\omega)$ is called the *frequency response* of the system. It is the value of the transfer function $H(s)$ evaluated at $s = j\omega$. It is also the Fourier transform of the impulse response $h(t)$, assuming that $h(t) = 0$ for $t < 0$.

For a general sinusoidal signal

$$x(t) = x_m \cos(\omega_0 t + \alpha), \tag{3.61}$$

a similar approach can be used to find $y(t)$. The steady-state output turns out to be given by an expression similar to (3.59)

$$y_{ss}(t) = M x_m \cos(\omega_0 t + \alpha + \phi), \tag{3.62}$$

In particular, if $\alpha = -\pi/2$, $x(t) = x_0 \sin(\omega_0 t)$ and $y_{ss}(t) = M x_0 \sin(\omega_0 t + \phi)$. The transient response is also similar, but the partial fraction expansion produces different coefficients for the exponentially decaying functions for different values of α.

3.4.3 Example: first-order system

Let

$$H(s) = \frac{k}{s+a}. \tag{3.63}$$

Case 1: The response to $x(t) = x_m \cos(\omega_0 t)$ is given by

$$
\begin{aligned}
Y(s) &= \frac{k}{s+a} x_m \frac{s}{s^2 + \omega_0^2} \\
&= \frac{-ak}{\omega_0^2 + a^2} x_m \frac{1}{s+a} + \frac{ak}{\omega_0^2 + a^2} x_m \frac{s}{s^2 + \omega_0^2} + \frac{\omega_0 k}{\omega_0^2 + a^2} x_m \frac{\omega_0}{s^2 + \omega_0^2}.
\end{aligned}
\tag{3.64}
$$

The first term of the response is the transient response, and corresponds to the time function

$$y_{tr}(t) = \frac{-ak}{\omega_0^2 + a^2} x_m e^{-at}. \tag{3.65}$$

The second and third terms constitute the steady-state response, which is given by

$$
\begin{aligned}
y_{ss}(t) &= \frac{ak}{\omega_0^2 + a^2} x_m \cos(\omega_0 t) + \frac{\omega_0 k}{\omega_0^2 + a^2} x_m \sin(\omega_0 t) \\
&= M x_m \cos(\omega_0 t + \phi),
\end{aligned}
\tag{3.66}
$$

with the magnitude and phase

$$M = \frac{k}{\sqrt{\omega_0^2 + a^2}} \qquad \phi = \angle(a - j\omega_0). \tag{3.67}$$

These quantities can be obtained directly from $H(j\omega_0) = k/(j\omega_0 + a)$ using (3.60). The magnitude decreases monotonically from the value of the DC gain k/a to zero, as ω varies from 0 to ∞. The phase decreases from 0 degrees to -90 degrees. The phase is also equal to

$$\phi = -\tan^{-1}\left(\frac{\omega_0}{a}\right), \tag{3.68}$$

but one must be careful to choose the correct quadrant for the inverse tangent function.

Case 2: For the response to $x(t) = x_m \sin(\omega_0 t)$, the partial fraction expansion gives

$$
\begin{aligned}
Y(s) &= \frac{k}{s+a} x_m \frac{\omega_0}{s^2 + \omega_0^2} \\
&= \frac{k\omega_0}{\omega_0^2 + a^2} x_m \frac{1}{s+a} - \frac{\omega_0 k}{\omega_0^2 + a^2} x_m \frac{s}{s^2 + \omega_0^2} + \frac{ak}{\omega_0^2 + a^2} x_m \frac{\omega_0}{s^2 + \omega_0^2},
\end{aligned}
\tag{3.69}
$$

so that the transient response is

$$y_{tr}(t) = \frac{k\omega_0}{\omega_0^2 + a^2} x_m e^{-at},$$ (3.70)

and the steady-state response is

$$
\begin{aligned}
y_{ss}(t) &= -\frac{\omega_0 k}{\omega_0^2 + a^2} x_m \cos(\omega_0 t) + \frac{ak}{\omega_0^2 + a^2} x_m \sin(\omega_0 t) \\
&= M x_m \sin(\omega_0 t + \phi).
\end{aligned}
$$ (3.71)

As noted earlier, the steady-state response is simply shifted if the input signal is shifted, but the transient response varies and is not simply shifted. For different phases of the input signal, there can be large variations in the transient response.
Case 3: The response to $x(t) = x_m \cos(\omega_0 t + \alpha)$ can be computed using

$$x_m \cos(\omega_0 t + \alpha) = x_m \cos(\omega_0 t) \cos(\alpha) - x_m \sin(\omega_0 t) \sin(\alpha).$$ (3.72)

Based on the linearity of the system, the response is a linear combination of the responses computed earlier, with

$$
\begin{aligned}
y_{tr}(t) &= \left(\frac{-ak}{\omega_0^2 + a^2} x_m e^{-at} \right) \cos(\alpha) - \left(\frac{k\omega_0}{\omega_0^2 + a^2} x_m e^{-at} \right) \sin(\alpha) \\
&= \left(\frac{k}{\omega_0^2 + a^2} x_m e^{-at} \right) (-a\cos(\alpha) - \omega_0 \sin(\alpha)).
\end{aligned}
$$ (3.73)

Similarly

$$y_{ss}(t) = M x_m \cos(\omega_0 t + \phi + \alpha),$$ (3.74)

as expected. It is possible to find values of α such that the transient response is maximized, as well as such that the transient response is zero. In an electrical circuit where an AC source is suddenly switched on, the transient currents may vary significantly depending on the time of switching.

3.5 Effect of initial conditions

In the analysis of the simple RL circuit, it was assumed that the current $i(t)$ was initially zero. A nonzero value could be accounted for, but would require an additional term in the response. The purpose of this section is to investigate the effect of such nonzero initial conditions in differential equations. We consider a general input-output differential equation

$$\frac{d^n y}{dt^n} + a_{n-1} \frac{d^{n-1} y}{dt^{n-1}} + \cdots + a_1 \frac{dy}{dt} + a_0 y = b_n \frac{d^n x}{dt^n} + b_{n-1} \frac{d^{n-1} x}{dt^{n-1}} + \cdots + b_1 \frac{dx}{dt} + b_0 x.$$ (3.75)

Without loss of generality, we let the first coefficient be equal to 1 (if needed, both sides can be divided by the first coefficient to get the result). For the Laplace transform analysis, recall that

$$y_1 = \frac{dy}{dt} \Rightarrow Y_1(s) = sY(s) - y(0). \tag{3.76}$$

Therefore

$$y_2 = \frac{d^2y}{dt^2} = \frac{dy_1}{dt} \Rightarrow Y_2(s) = sY_1(s) - y_1(0) = s^2Y(s) - sy(0) - \dot{y}(0), \tag{3.77}$$

where we used the notation

$$\frac{dy}{dt} = \dot{y} \quad \text{and} \quad y_1(0) = \dot{y}(0) = \left[\frac{dy}{dt}\right]_{t=0}. \tag{3.78}$$

The procedure can be extended to higher-order derivatives.

Consider the case of a second-order input/output differential equation

$$\frac{d^2y}{dt^2} + a_1\frac{dy}{dt} + a_0y = b_2\frac{d^2x}{dt^2} + b_1\frac{dx}{dt} + b_0x. \tag{3.79}$$

Application of the Laplace transform to both sides yields

$$\begin{aligned}
s^2Y(s) \quad - \quad & sy(0) - \dot{y}(0) + a_1sY(s) - a_1y(0) + a_0Y(s) \\
= \quad & b_2s^2X(s) - b_2sx(0) - b_2\dot{x}(0) \\
& +b_1sX(s) - b_1x(0) + b_0X(s).
\end{aligned} \tag{3.80}$$

Note the distinction between $X(s)$, which is the Laplace transform of $x(t)$, and $x(0)$, which is the initial value of $x(t)$. The transform $Y(s)$ may be deduced to be

$$Y(s) = \underbrace{\frac{sy(0) + \dot{y}(0) + a_1y(0) - b_2sx(0) - b_2\dot{x}(0) - b_1x(0)}{s^2 + a_1s + a_0}}_{\substack{\text{Response to the initial conditions} \\ \text{or zero-input response } Y_{zi}(s)}}$$

$$+ \underbrace{\underbrace{\frac{b_2s^2 + b_1s + b_0}{s^2 + a_1s + a_0}}_{H(s) \text{ (transfer function)}} X(s)}_{\substack{\text{Response to the input} \\ \text{or zero-state response } Y_{zs}(s)}}. \tag{3.81}$$

The first term is the *response to the initial conditions*, and is also called the *zero-input response* $Y_{zi}(s)$. The second term is the *response to the input*, and is also called the *zero-state response* $Y_{zs}(s)$. It is the product of the transfer function $H(s)$ with the transform of the input signal. The following observations may be made, which also apply to systems of higher order:

- the response is the sum of the response due to the input and the response due to the initial conditions. The two components are independent. The response to the input is the response that is obtained for the same input but zero initial conditions, and the response to the initial conditions is the response that is obtained for the same initial conditions but zero input.

- the initial conditions are composed of the values of the input and output variables as well as their derivatives at $t = 0$. All the relevant history of the system for $t < 0$ is contained in those values, which may be viewed as the *state* of the system at $t = 0$.

- the transfer function and the response to initial conditions are rational functions of s with the same denominators. As a consequence, the response to the initial conditions is similar to the transient response defined earlier for step and sinusoidal inputs. Both can be lumped together as a "total" transient response. Boundedness of this part of the response is associated with the concept of *internal stability*.

- the response to the initial conditions converges to zero if the poles of the system are in the open left half-plane. This property is referred to as *asymptotic stability*. The condition for asymptotic stability is the same as for BIBO stability.

- the response to the initial conditions is bounded if the poles of the system are in the open left half-plane, or are non-repeated poles on the $j\omega-$axis. This property is sometimes referred to as *marginal stability*. However, a marginally stable system is *unstable* from the BIBO point of view.

- singular problems are encountered if a pole/zero cancellation occurs in the rational functions. For example, the response to some initial conditions may be bounded even if the transfer function has poles in the open right half-plane. Conversely, unstable poles may be cancelled in the transfer function, resulting in a BIBO stable system. In that case, however, the response to initial conditions may still be unbounded. Generally, cancellations of undesirable poles are viewed as conditions to be avoided, because of initial conditions and because cancellations can never be made exact in practice.

In summary, the definitions are:

Asymptotically stable	$y_{zi}(t) \to 0$ as $t \to \infty$
Marginally stable	$y_{zi}(t)$ bounded
Internally unstable	$y_{zi}(t)$ unbounded

while the tests on the poles are

Asymptotically stable	All poles with $\text{Re}(s) < 0$
Marginally stable	All poles with $\text{Re}(s) < 0$ + possible non-repeated poles on the $j\omega-$axis
Internally unstable	At least one pole with $\text{Re}(s) > 0$ or repeated pole on the $j\omega-$axis

Asymptotic stability is equivalent to bounded-input bounded-output stability.

3.6 State-space representations

3.6.1 Example of a state-space model

State-space models are similar to input/output differential equations. However, they arise more naturally from the modelling of physical systems. For example, consider the RLC circuit shown in Fig. 3.18. Using standard circuit analysis tools, the following equations may be written

$$
\begin{aligned}
u &= L\frac{dx_1}{dt} + x_2 + y \\
x_1 &= C\frac{dx_2}{dt} \\
y &= Rx_1.
\end{aligned}
\tag{3.82}
$$

Note that the voltage $u(t)$ is not a step function here, but the input to the system, in accordance with standard state-space notation.

The two differential equations for the circuit may be written as

$$
\begin{aligned}
\frac{dx_1}{dt} &= -\frac{R}{L}x_1 - \frac{1}{L}x_2 + \frac{1}{L}u \\
\frac{dx_2}{dt} &= \frac{1}{C}x_1,
\end{aligned}
\tag{3.83}
$$

or, in matrix form,

$$
\begin{pmatrix} dx_1/dt \\ dx_2/dt \end{pmatrix} = \begin{pmatrix} -R/L & -1/L \\ 1/C & 0 \end{pmatrix} \begin{pmatrix} x_1 \\ x_2 \end{pmatrix} + \begin{pmatrix} 1/L \\ 0 \end{pmatrix} u,
\tag{3.84}
$$

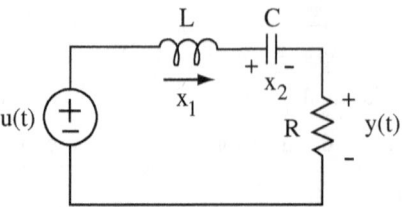

Figure 3.18: RLC circuit

and the output of the system is

$$y = \begin{pmatrix} R & 0 \end{pmatrix} \begin{pmatrix} x_1 \\ x_2 \end{pmatrix}. \tag{3.85}$$

Equations (3.84) and (3.85) constitute a state-space model for the RLC circuit.

3.6.2   General form of a state-space model

In general, a state-space representation has the form

$$\begin{aligned} \frac{dx}{dt} &= Ax + Bu \\ y &= Cx + Du, \end{aligned} \tag{3.86}$$

where:

x is a column vector of dimension n, called the *state vector*

u is a scalar signal, and the input of the system

y is a scalar signal, and the output of the system

A is an $n \times n$ matrix, B is a column vector of dimension n

C is a row vector of dimension n, and D is a scalar

The dimensions of all the elements and of the products are shown in Fig. 3.19. Generally, dx/dt is denoted $\dot{x}$.

To obtain a state-space model for circuits, a systematic procedure consists in:

1. defining a state vector with the voltages on the capacitors and the currents in the inductors as components.

2. writing equations using Kirchhoff's voltage law, Kirchhoff's current law, and element descriptions.

3. converting the equations to state-space form.

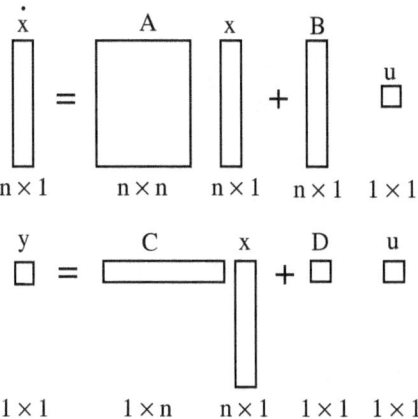

Figure 3.19: Matrix products in the state-space model

Many other physical systems can also be represented with state-space models, using standard modelling techniques.

3.6.3   State-space analysis

Some simple yet general results may be obtained by applying the Laplace transform to the state-space model. The equations for the state-space model are

$$\dot{x} = Ax + Bu$$
$$y = Cx + Du, \tag{3.87}$$

so that, in the s-domain

$$sX(s) - x(0) = AX(s) + BU(s)$$
$$Y(s) = CX(s) + DU(s). \tag{3.88}$$

The first (vector) equation gives

$$sX(s) - AX(s) = (sI - A)X(s) = x(0) + BU(s), \tag{3.89}$$

where I is the identity matrix with dimension $n \times n$. The transform of the output is therefore

$$Y(s) = \underbrace{C(sI - A)^{-1}x(0)}_{\substack{\text{Response to initial conditions} \\ \text{or zero-input response } Y_{zi}(s)}} + \underbrace{\left(C(sI - A)^{-1}B + D\right)U(s)}_{\substack{\text{Response to input} \\ \text{or zero-state response } Y_{zs}(s)}}.$$

(3.90)

The dimensions of the terms in the expression are, in the order in which they appear

$$\begin{aligned}
1 \times 1 &= (1 \times n) \times (n \times n) \times (n \times 1) \\
&\quad + ((1 \times n) \times (n \times n) \times (n \times 1) + (1 \times 1)) \times (1 \times 1).
\end{aligned}$$
(3.91)

As in the case of input/output differential equations, the output $Y(s)$ obtained from the state-space model has two terms: a term due to initial conditions, and a term due to the input. The transfer function is found to be

$$H(s) = C(sI - A)^{-1}B + D.$$
(3.92)

Although this transfer function may be complicated to compute, the poles are determined by a simple equation related to the denominator of the matrix $(sI - A)^{-1}$. Specifically, the denominator is $\det(sI - A)$ and, therefore, the poles are given by the roots of $\det(sI - A) = 0$. These roots are called the eigenvalues of the matrix A, and may be computed using a mathematical software package. **Example:** for the RLC circuit,

$$A = \begin{pmatrix} -R/L & -1/L \\ 1/C & 0 \end{pmatrix}, \quad B = \begin{pmatrix} 1/L \\ 0 \end{pmatrix}, \quad C = (\,R \;\; 0\,), \quad D = 0.$$
(3.93)

Using the formula for the inverse of a 2×2 matrix

$$M^{-1} = \begin{pmatrix} M_{11} & M_{12} \\ M_{21} & M_{22} \end{pmatrix}^{-1} = \frac{1}{M_{11}M_{22} - M_{21}M_{12}} \begin{pmatrix} M_{22} & -M_{12} \\ -M_{21} & M_{11} \end{pmatrix},$$
(3.94)

one finds that the transfer function is

$$\begin{aligned}
H(s) &= (\,R \;\; 0\,) \begin{pmatrix} s + R/L & 1/L \\ -1/C & s \end{pmatrix}^{-1} \begin{pmatrix} 1/L \\ 0 \end{pmatrix} \\
&= \frac{1}{s^2 + (R/L)s + 1/LC} (\,R \;\; 0\,) \begin{pmatrix} s & -1/L \\ 1/C & s + R/L \end{pmatrix} \begin{pmatrix} 1/L \\ 0 \end{pmatrix} \\
&= \frac{(R/L)s}{s^2 + (R/L)s + 1/LC}.
\end{aligned}$$
(3.95)

Note that the elements of the circuit are connected in series, so that the impedance of the circuit is

$$Z(s) = \frac{V(s)}{I(s)} = sL + \frac{1}{sC} + R = \frac{s^2 LC + 1 + sCR}{sC}, \tag{3.96}$$

so that the transfer function can be computed independently to be

$$H(s) = \frac{RI(s)}{V(s)} = \frac{sRC}{s^2 LC + sCR + 1}, \tag{3.97}$$

which is the same result. Note that there is a zero in the transfer function at $s = 0$, because of the blocking of DC signals by the capacitor.

The response to initial conditions is

$$C(sI - A)^{-1} x(0) = \frac{R(sx_1(0) - (1/L)x_2(0))}{s^2 + (R/L)s + 1/LC}, \tag{3.98}$$

where $x_1(0)$ is the current in the inductor and $x_2(0)$ is the voltage on the capacitor, both at $t = 0$. The denominator in the expressions is $\det(sI - A) = s^2 + (R/L)s + 1/LC$, and the roots of the polynomial are the poles of the system.

3.6.4   State-space realizations

An interesting fact is that a state-space model can always be created that "realizes" a transfer function $H(s) = N(s)/D(s)$, provided that $\deg N(s) \leqslant \deg D(s)$. The procedure works as follows. First, use polynomial division to divide $N(s)$ by $D(s)$, so that $N(s) = q_0 D(s) + R(s)$, with $\deg R(s) < \deg D(s)$, and the quotient q_0 is a scalar. Then, denoting

$$\frac{R(s)}{D(s)} = \frac{b_{n-1} s^{n-1} + \cdots + b_1 s + b_0}{s^n + a_{n-1} s^{n-1} + \cdots + a_1 s + a_0}, \tag{3.99}$$

the matrices of a state-space realization are

$$A = \begin{pmatrix} 0 & 1 & 0 & \cdots & 0 \\ 0 & 0 & 1 & 0 & \cdots & 0 \\ \vdots & & & & \\ 0 & \cdots & & \cdots & 0 & 1 \\ -a_0 & -a_1 & & \cdots & & -a_{n-1} \end{pmatrix}, \quad B = \begin{pmatrix} 0 \\ \vdots \\ 0 \\ 1 \end{pmatrix}$$

$$C = \begin{pmatrix} b_0 & \cdots & b_{n-1} \end{pmatrix}, \quad D = q_0. \tag{3.100}$$

This state-space realization can be implemented using integrators, multipliers, and summing junctions, using the diagram of Fig. 3.20.

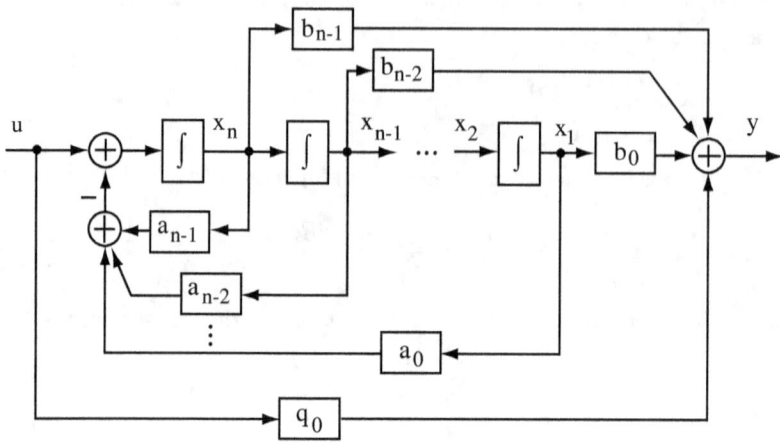

Figure 3.20: Realization of a linear time-invariant system using integrators

To prove that the state-space system indeed has the required transfer function, note that the equations of the system are

$$\dot{x}_1 = x_2, \ \dot{x}_2 = x_3, \ \cdots, \ \dot{x}_{n-1} = x_n$$
$$\dot{x}_n = -a_0 x_1 \cdots - a_{n-1} x_n + u$$
$$y = b_0 x_1 \cdots + b_{n-1} x_n + q_0 u, \tag{3.101}$$

so that, applying the Laplace transform,

$$X_2(s) = sX_1(s), ..., \ X_n(s) = s^{n-1} X_1(s), \tag{3.102}$$

and

$$X_1(s) = \frac{1}{s^n + a_{n-1} s^{n-1} + \cdots + a_1 s + a_0} U(s). \tag{3.103}$$

Therefore

$$
\begin{aligned}
Y(s) &= \frac{b_{n-1} s^{n-1} + \cdots + b_1 s + b_0}{s^n + a_{n-1} s^{n-1} + \cdots + a_1 s + a_0} U(s) + q_0 U(s) \\
&= \left(\frac{R(s)}{D(s)} + q_0 \right) U(s) \\
&= \frac{N(s)}{D(s)} U(s), \tag{3.104}
\end{aligned}
$$

which is the desired result.

For example, the transfer function of the RLC circuit considered earlier

$$H(s) = \frac{(R/L)s}{s^2 + (R/L)s + 1/LC},$$
(3.105)

may be realized using

$$A = \begin{pmatrix} 0 & 1 \\ -1/LC & -R/L \end{pmatrix}, \quad B = \begin{pmatrix} 0 \\ 1 \end{pmatrix}, \quad C = \begin{pmatrix} 0 & R/L \end{pmatrix}, \quad D = 0.$$
(3.106)

Note that this state-space representation is different from the one that gave rise to the transfer function. Indeed, a state-space model is not unique. For a given state-space model, another model can be obtained by applying what is called a *similarity transformation*. A new state z is defined through

$$z = Px,$$
(3.107)

where P is an invertible matrix. Then

$$\begin{aligned} \dot{z} &= P\dot{x} = PAx + PBu = PAP^{-1}z + PBu \\ y &= Cx + Du = CP^{-1}z + Du, \end{aligned}$$
(3.108)

so that z is the state of a model with matrices $\bar{A} = PAP^{-1}, \bar{B} = PB, \bar{C} = CP^{-1}$, $\bar{D} = D$, and identical transfer function. It can be shown that if two state-space realizations implement the same transfer function *without pole/zero cancellations* in $H(s)$, they are related by a similarity transformation.

Realizability: to be *realizable* as a state-space system, a transfer function must be proper. A non-proper transfer function is realizable as an input-ouput differential equation, but is problematic in practice because the magnitude of the frequency response is unbounded for increasing frequencies. Filtering may be added to approximate the system as a proper transfer function. For example, $H(s) = s$ is implemented as $s\,a/(s+a)$ where $a > 0$ is large. The approximate transfer function can be implemented as a state-space system.

3.7 Problems

Problem 3.1: A model of a brush DC motor without load is

$$\begin{aligned} L\frac{di}{dt} &= v - Ri - K\omega \\ J\frac{d\omega}{dt} &= Ki, \end{aligned}$$
(3.109)

where R (Ω) is the rotor resistance, L (H) is the rotor inductance, K (N m/A or V s) is the motor torque constant (also the back-emf constant), J (kg m^2) is the inertia of the motor. The input of the system is the voltage v (V) applied to the motor and the output is the rotor velocity ω (rad/s). The current i (A) is considered to be an internal variable (or "state").

(a) Find the transfer function from v to ω. Give the values of the poles and zeros of the transfer function.

(b) Find the approximate transfer function that is obtained when $L = 0$, and give the values of its poles and zeros.

(c) Compare the numerical values obtained for parts (a) and (b) when $R = 0.7\,\Omega$, $L = 2.5\ 10^{-3}$ H, $K = 0.07$ N m/A, and $J = 5.7\ 10^{-5}$ kg m^2.

Problem 3.2: (a) Find the transfer function from $X(s)$ to $Y(s)$ for the system shown in Fig. 3.21.

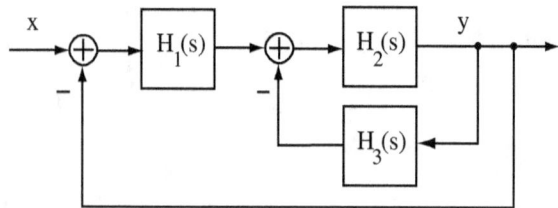

Figure 3.21: System for problem 3.2, part (a)

(b) Repeat part (a) for the system shown in Fig. 3.22.

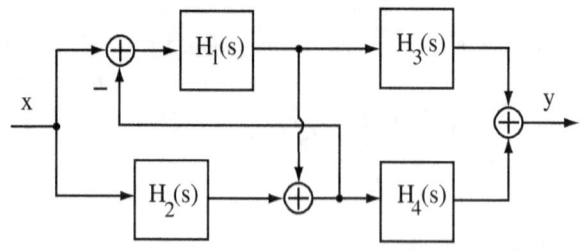

Figure 3.22: System for problem 3.2, part (b)

Problem 3.3: Determine which transfer functions are stable.

(a) $H(s) = \dfrac{s-1}{(s+2)^2}$

(b) $H(s) = \dfrac{1}{(s^2 + 4)}$

(c) $H(s) = \dfrac{s}{(s + 3)^2}$

(d) $H(s) = \dfrac{s}{(s^2 - 4)}$

(e) $H(s) = \dfrac{1}{s(s + 1)}$

(f) $H(s) = \dfrac{1}{s^2(s + 1)}$

For the unstable systems, give an example of a bounded input that yields an unbounded output.

Problem 3.4: For the systems given below, calculate the DC gain. Then, calculate the step responses using partial fraction expansions and compare the steady-state values to the values predicted by the DC gain.

(a) $H(s) = \dfrac{2}{s^2 + 2s + 1}$

(b) $H(s) = \dfrac{-s - 2}{s^2 + 2s + 2}$

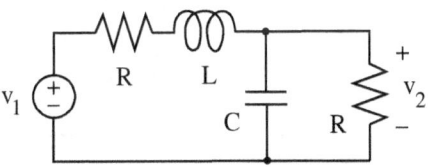

Figure 3.23: Circuit for problem 3.5

Problem 3.5: Consider the circuit of Fig. 3.23.

All the initial conditions in the circuit are zero.

(a) Using complex impedances, calculate the transfer function from v_1 to v_2. Show that $H(s) = 1/(s^2 + 2s + 2)$ when $R = L = C = 1$.

(b) Using partial fraction expansions, calculate the response $v_2(t)$ to $v_1(t) = 5\cos(t)$. Indicate which part of the response is the transient response and which part is the steady-state response.

(c) Calculate the steady-state response using the frequency response, and compare the results to those of part (b).

(d) Is there a frequency ω such that the steady-state response to $\cos(\omega t)$ is zero for all t ? Is there a frequency ω such that the steady-state response to $\cos(\omega t)$ is proportional to $\sin(\omega t)$?

Problem 3.6: (a) Write a state-space description for the circuit of problem 3.5.

(b) Calculate the transfer function using $H(s) = C(sI - A)^{-1}B + D$ and compare the result to the result obtained in problem 3.5.

(c) Calculate the additional response due to an initial current in the inductor and an initial voltage on the capacitor using $C(sI - A)^{-1}x(0)$. Give an estimate of the time required for the transient voltage to decay to negligible values when $R = L = C = 1$.

Problem 3.7: (a) Calculate the Laplace transform $Y(s)$ of the solution of the differential equation

$$\frac{dy^2(t)}{dt^2} + 4\frac{dy(t)}{dt} + 29y(t) = \cos(t), \tag{3.110}$$

with initial conditions $y(0)$, $\dot{y}(0)$.

(b) Without performing a partial fraction expansion, sketch the response to the initial conditions (*i.e.*, the zero-input response).

Problem 3.8: Write a state-space description for the circuit shown in Fig. 3.24

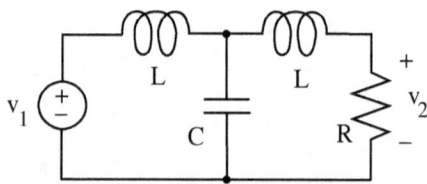

Figure 3.24: Circuit for problem 3.8

The input is v_1 and the output is v_2. Give the equation that must be solved to find the poles of the system and solve it for $R = L = C = 1$.

Problem 3.9: (a) Find the response $y(t)$ of the system with transfer function $H(s) = \frac{1}{s(s+1)}$ and input $x(t) = 1$.

(b) Find the response $y(t)$ of the system with transfer function $H(s) = \frac{1}{s(s+1)}$ and input $x(t) = \sin(t)$.

(c) Is a system BIBO stable if its response to a step input is bounded?

(d) Is a system BIBO unstable if its response to a step input is unbounded?

(e) Is the response of $H(s) = \frac{1}{(s+1)^2(s+4)}$ to $X(s) = \frac{s+3}{s(s+4)}$ bounded? Does it converge to a steady-state value? If so, to what value?

(f) Is the response of $H(s) = \frac{1}{(s+1)^2(s+4)}$ to $X(s) = \frac{1}{(s^2+1)^2}$ bounded? Does it converge to a steady-state value? If so, to what value?

Problem 3.10: Find the transfer function $H(s) = Y(s)/X(s)$ for the system of Fig. 3.25.

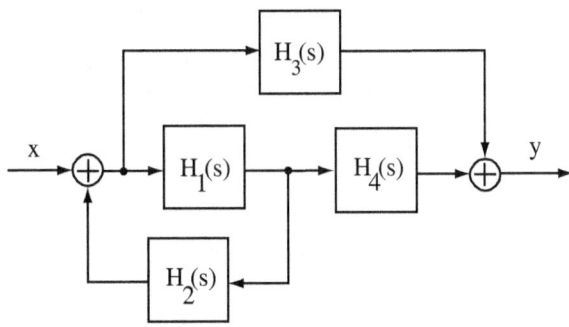

Figure 3.25: System for problem 3.10

Problem 3.11: (a) Give the transform $Y(s)$ for the system of Fig. 3.26. Assume that the input has transform $U(s)$, and that the integrators have initial conditions $x_1(0)$ and $x_2(0)$.

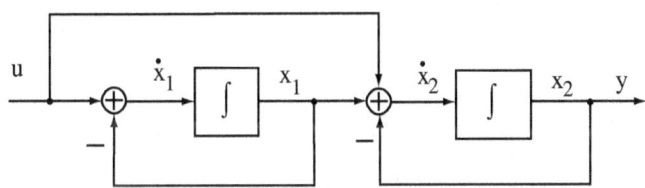

Figure 3.26: System for problem 3.11

(b) Give the steady-state value of the output when the input $u(t) = 5$ for all t (if the limit does not exist, indicate why).

Problem 3.12: (a) Find the transfer function $W(s)/R(s)$ for the system of Fig. 3.27.

(b) Given that the steady-state response of $P(s)$ is $y_{ss}(t) = \sin(10t - 60°)$ when $x(t) = \sin(10t)$, what is the steady-state response $w_{ss}(t)$ of the system of part (a) when $r(t) = \sin(10t)$? Give the condition that $P(s)$ must satisfy for the result to be true.

Problem 3.13: (a) Write a state-space model for the circuit of Fig. 3.28 and, using the A matrix, give the polynomial whose roots are the poles of the system.

(b) Using the method of your choice (standard circuit calculations using complex impedances are recommended), obtain the transfer function from u to y for the circuit of part (a). Give the DC gain of the system and the location of the

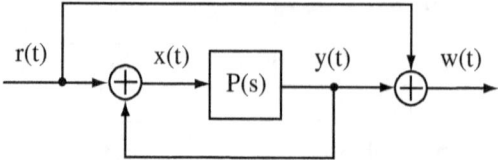

Figure 3.27: System for problem 3.12

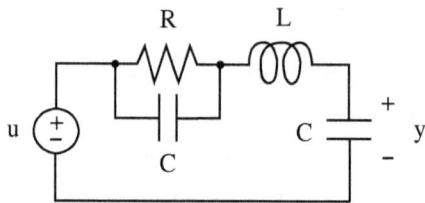

Figure 3.28: Circuit for problem 3.13

zero(s).

Problem 3.14: Calculate the transfer function $P(s) = Y(s)/X(s)$ for the system shown in Fig. 3.29.

Problem 3.15: (a) Calculate the transient response $y_{tr}(t)$ associated with the output $y(t)$ of a system with transfer function

$$P(s) = \frac{s}{(s+1)^2} \tag{3.111}$$

and input $x(t) = 2\cos(t)$ (the steady-state response is not needed).

(b) Without performing a partial fraction expansion, give the value of the steady-state output $y_{ss}(t)$ of the system of part (a) for a constant input $x(t) = 2$. If the steady-state output does not exist, explain why.

(c) Without performing a partial fraction expansion, give the value of the steady-state response of the system of part (a) to an input $x(t) = 5\cos(2t)$. If the steady-state response does not exist, explain why.

(d) A system with transfer function $P_1(s)$ has steady-state response $2\cos(t-30°)$ when an input $x(t) = \cos(t)$ is applied to the system. Another system with transfer function $P_2(s)$ has steady-state response $3\sin(t + 60°)$ when an input $x(t) = \sin(t)$ is applied. What is the steady-state response of the cascade system $P(s) = P_1(s)P_2(s)$ if an input $x(t) = \cos(t)$ is applied to the system?

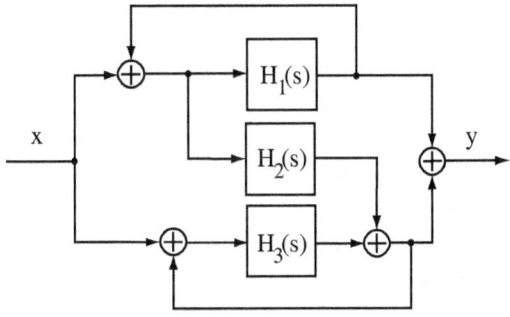

Figure 3.29: System for problem 3.14

Problem 3.16: (a) For the circuit of Fig. 3.30, calculate the voltage $v_2(t)$ that is observed for $t \geqslant 0$ if $v_1(t) = 15$ V, $R = 10 \, \Omega$, $C = 100 \, \mu$F, and the initial voltage on the capacitor is $v_c(0) = 5$ V. Sketch the voltage $v_2(t)$, being careful to label the axes precisely.

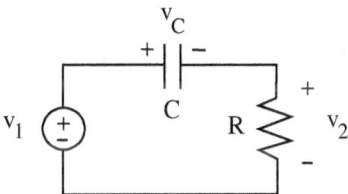

Figure 3 30: Circuit for problem 3.16 (a)

(b) What are the poles of a system whose state-space representation is such that:

- the matrix $A = 0$,

- the matrix $A = kI$ (the identity matrix multiplied by a constant k).

Problem 3.17: (a) Indicate whether the following system is BIBO stable.

$$H(s) = \frac{s - 1}{(s + 2 + j)^2 (s + 2 - j)^2}.$$ (3.112)

(b) Indicate whether the following system is BIBO stable.

$$H(s) = \frac{s + 1}{(s + j)(s - j)}.$$ (3.113)

(c) Indicate whether the signal with the following transform is bounded, and whether it converges.

$$X(s) = \frac{s^2 + 9}{(s^2 - 1)(s^2 + 4)}. \tag{3.114}$$

(d) A signal $x(t) = \cos(4t)$ is applied to a system with transfer function $H(s) = 1/(s^2 + 4)$. Is the output bounded?

(e) A signal that converges to zero is applied to a BIBO stable system. The transform of the signal and the transfer function of the system are proper, rational functions of s. Indicate whether the output of the system: (1) is bounded, (2) converges, and (3) converges to zero.

Problem 3.18: (a) Find the transfer function $H(s) = Y(s)/X(s)$ for the system of Fig. 3.31.

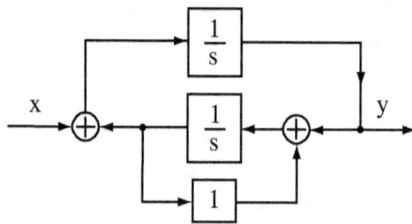

Figure 3.31: System for problem 3.18

(b) What can you say about the steady-state value of the output $y_{ss}(t)$ for a constant input $x(t) = 5$?

Problem 3.19: (a) *Using the frequency response*, determine what the steady-state output $y_{ss}(t)$ is for an input $x(t) = \sin(t)$ and a system with transfer function

$$H(s) = \frac{-(s - 1)}{(s + 1)^2}. \tag{3.115}$$

(b) For the system and input signal of part (a), use partial fraction expansions to determine the transient response $y_{tr}(t)$ (only the transient response is needed).

Problem 3.20: Write a state-space model for the circuit of Fig. 3.32.

Problem 3.21: (a) Consider the system with input x and output y described by the differential equation

$$\frac{d^2 y}{dt^2} = ay + bx. \tag{3.116}$$

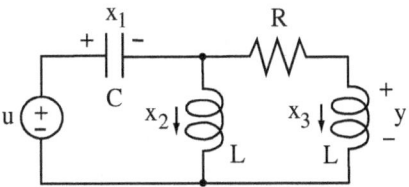

Figure 3.32: Circuit for problem 3.20

Give the Laplace transform of the response to initial conditions, or zero-input response $Y_{zi}(s)$.

(b) Give the time-domain function $y_{zi}(t)$ for $a = -1$ and for $a = 1$.

Problem 3.22: (a) By means of the Laplace transform, find the solutions $x_1(t)$ and $x_2(t)$ of the following system

$$\frac{dx_1(t)}{dt} = x_2$$
$$\frac{dx_2(t)}{dt} = -2x_1 - 3x_2 + u(t), \qquad (3.117)$$

where $x_1(0) = 1$, $x_2(0) = 0$, and $u(t)$ is a step input of magnitude 1.

(b) Write a state-space realization for the system in Fig. 3.33 and give the values of the poles of the system.

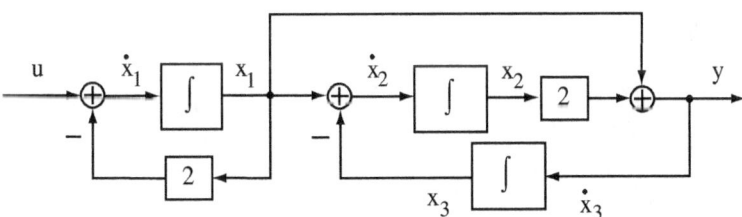

Figure 3.33: System for problem 3.22 (b)

Chapter 4

Stability and performance of control systems

4.1 Control system characteristics

A standard control system is shown in Fig. 4.1. The following elements may be recognized:

- $P(s)$ is the plant, or system to be controlled

- $C(s)$ is the controller, compensator, or control system

- y is the plant output

- u is the control input

- r is the reference input

- e is the tracking error

The objective of control system design is to find a compensator such that the plant output matches the reference input as closely as possible. Then, the tracking error will be zero, or close to zero. The reference input may be specified by a human operator or computed automatically by a computer.

Practically, most plants are physical systems and are best described by continuous-time models. On the other hand, control systems are typically computer-based, and operate in a discrete-time domain associated with a certain sampling frequency. Sensors must be chosen to provide the desired measurements of the plant output(s), and actuator(s) must provide the required power to drive the system. Analog to digital (A/D) and digital to analog (D/A) converters transform the signals between the continuous-time and discrete-time domains. Overall, the physical structure of a control system is shown in Fig. 4.2.

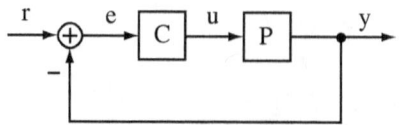

Figure 4.1: Basic control system

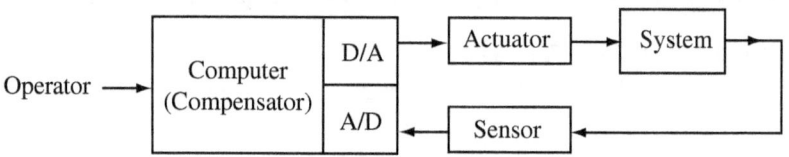

Figure 4.2: Practical implementation of control system

The objectives of the control system are to:

- maintain the stability of the system (either by preserving the inherent stability of a system or by stabilizing the system if it is unstable).

- ensure the tracking of reference inputs (including convergence to steady-state reference values as well as fast and smooth transient response to varying inputs).

- reject disturbances affecting the system.

- be sufficiently insensitive to plant uncertainties and time variations.

- tolerate the presence of noise on the measurements.

One distinguishes between *feedforward* and *feedback* control systems (both shown in Fig. 4.3). A feedfoward control system typically implements an approximation of the inverse of the plant. Generally, the advantages of a feedforward control system are that stability is easier to maintain, and that no output sensor is needed (hence, the system is insensitive to measurement noise). However, feedback systems are typically preferred because the effect of disturbances may be reduced, and plant variations or uncertainties may be compensated for. Feedback systems are also the only option available for unstable plants. In practice, control systems are often a blend of feedforward and feedback control.

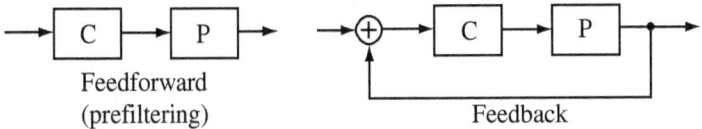

Figure 4.3: Feedforward and feedback control

For stability and for an adequate transient response, the desired locations of closed-loop poles are shown in Fig. 4.4. The poles should be sufficiently far in the left half-plane that the response of the system is fast, and sufficiently close to the real axis that the responses do not exhibit large transient oscillations.

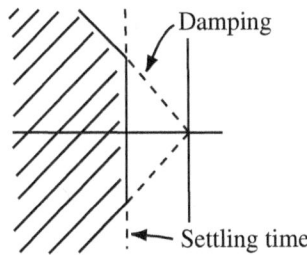

Figure 4.4: Region for desirable pole locations in the s-plane

4.2   Proportional control

A simple controller is the *proportional control* system

$$C(s) = k_P, \tag{4.1}$$

where $k_P > 0$ is a fixed gain. The feedback system is shown on Fig. 4.5 for a plant

$$P(s) = \frac{1}{s+1}. \tag{4.2}$$

A feedforward gain k_0 was added that will be discussed shortly.

The closed-loop transfer function is given by

$$P_{CL}(s) = \frac{Y(s)}{R(s)} = \frac{k_0 k_P}{s + 1 + k_P}. \tag{4.3}$$

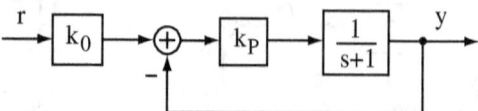

Figure 4.5: Proportional control of a first-order plant

The transfer function has a single pole at

$$s = -1 - k_P. \tag{4.4}$$

The system is stable for all $k_P > 0$, and the DC gain of the transfer function is

$$P_{CL}(0) = \frac{k_0 k_P}{1 + k_P}. \tag{4.5}$$

The feedforward gain was inserted in Fig. 4.5 so that the DC gain could be made equal to 1 by setting

$$k_0 = \frac{1 + k_P}{k_P}. \tag{4.6}$$

Then, the output will track a constant reference input in the steady-state.

In closed-loop, the original pole at $s = -1$ moves to an arbitrary value determined by k_P. The closed-loop system response can be made to respond faster. For example, for $k_P = 1$, $k_0 = 2$,

$$\frac{Y(s)}{R(s} = \frac{2}{s+2}, \tag{4.7}$$

which means that the closed-loop system responds twice as fast as the original system. The input signal is

$$U(s) = \frac{2(s+1)}{s+2} R(s). \tag{4.8}$$

The signal is the same as if a feedforward controller was used to cancel the pole of the plant and replace it by a faster pole. However, there are fundamental differences between moving a pole by pole/zero cancellation and by using feedback.

For a unit step input $R(s) = 1/s$,

$$U(s) = \frac{1}{s} + \frac{1}{s+2} \Leftrightarrow u(t) = 1 + e^{-2t}. \tag{4.9}$$

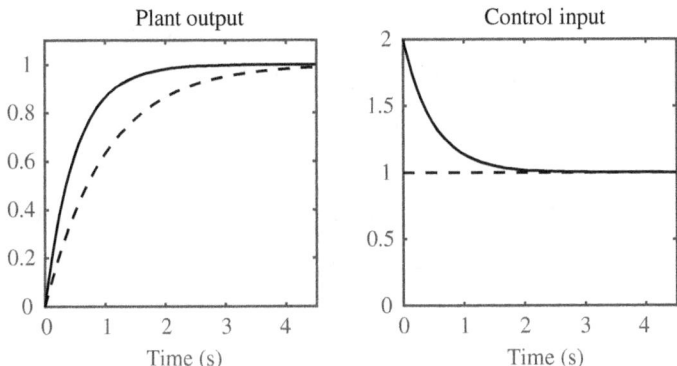

Figure 4.6: Plant output (left) and control input (right) for the open-loop system (dashed) and with proportional feedback (solid)

The response of the system and the input signal are shown on Fig. 4.6 for the open-loop and closed-loop systems. One finds that the response is accelerated, although the result is obtained by applying a much larger input signal at the beginning of the response.

In practice, a pole cannot be moved arbitrarily far in the left half-plane due to input constraints, as well as other factors. However, feedback is helpful to increase the speed of response within limits, to improve damping, or to stabilize a system. Fig. 4.5 required the addition of a feedforward gain to achieve tracking. An alternative and often preferable solution consists in using a controller with a pole at $s = 0$.

4.3 Steady-state error and integral control

4.3.1 Tracking of constant reference inputs

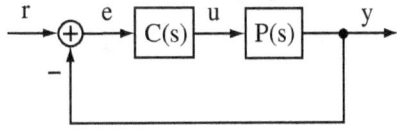

Figure 4.7: Feedback control system

A standard control system is shown in Fig. 4.7, where we assumed that the plant and control systems are linear time-invariant and described by transfer functions $P(s)$ and $C(s)$, respectively. A significant signal in this system is the *tracking error*

$$e(t) = r(t) - y(t), \tag{4.10}$$

which is expected to remain close to zero. The *steady-state error* is defined to be

$$e_{ss} = \lim_{t \to \infty} e(t), \tag{4.11}$$

assuming that the limit exists. The reference input $r(t)$ is taken to be constant, *i.e.*,

$$r(t) = r_m, \quad R(s) = \frac{r_m}{s}. \tag{4.12}$$

It is typical for the reference input of a control system to be constant for relatively long periods of time and for the steady-state error to be a significant consideration. The infinite-time limit is really an approximation of time periods that are long compared to the transient response of the system. For example, the reference speed of a cruise control system may remain constant for minutes, with the speed itself converging within a few seconds.

The Laplace transform may be used to analyze the feedback system of Fig. 4.7. The tracking error is given by

$$E(s) = R(s) - Y(s) = R(s) - P(s)C(s)E(s), \tag{4.13}$$

so that

$$E(s) = \frac{1}{1 + P(s)C(s)} R(s). \tag{4.14}$$

According to the analysis of step responses, the steady-state error for a constant reference input is

$$e_{ss} = \underbrace{\frac{1}{1 + P(0)C(0)}}_{\text{DC gain of the transfer function from } r \to e} r_m. \tag{4.15}$$

Recall that the closed-loop system must be stable for the steady-state error to be well-defined.

Alternatively, one could calculate $Y(s)$ using

$$Y(s) = \frac{P(s)C(s)}{1 + P(s)C(s)} R(s) \tag{4.16}$$

and the steady-state plant output would be

$$\lim_{t\to\infty} y(t) = \underbrace{\frac{P(0)C(0)}{1 + P(0)C(0)}}_{\text{DC gain of the transfer function from } r \to y} r_m. \tag{4.17}$$

For the output signal to converge to the reference input, the DC gain of the transfer function from r to y should be equal to 1. It is easy to check that, since $e_{ss} = r_m - y_{ss}$, the result is the same as the one obtained with (4.15).

Next, we define *perfect tracking*, as the condition in which $e_{ss} = 0$, or $y_{ss} = r_m$. Let

$$P(s) = \frac{n_p(s)}{d_p(s)}, \quad C(s) = \frac{n_c(s)}{d_c(s)}, \tag{4.18}$$

where $n_p(s)$, $d_p(s)$, $n_c(s)$, and $d_c(s)$ are polynomials. The condition for perfect tracking is that

$$\frac{1}{1 + P(0)C(0)} = \frac{d_p(0)d_c(0)}{n_p(0)n_c(0) + d_p(0)d_c(0)} = 0. \tag{4.19}$$

Thus, the condition is satisfied if and only if

$$d_p(0) = 0 \text{ or } d_c(0) = 0. \tag{4.20}$$

In other words, **perfect tracking of constant reference inputs** is achieved if:

- the closed-loop system is stable.

- either $P(s)$ has a pole at $s = 0$ or $C(s)$ has a pole at $s = 0$

There is also a technical requirement that neither $P(s)$ nor $C(s)$ have a zero at $s = 0$. Obviously, if the plant has a zero at $s = 0$, *i.e.*, has zero DC gain, there is no possibility of tracking constant reference inputs.

4.3.2 Rejection of constant disturbances

The problem of rejecting constant disturbances turns out to be similar, although not identical, to the problem of perfect tracking. Constant disturbances may be caused by wind in the case of an autopilot for aircraft, or by the slope of the road in the case of a cruise control system. A similar problem formulation may be posed as for tracking, assuming that the disturbance is $d(t) = d_m$, or $D(s) = d_m/s$. We assume that the disturbance is added at the input of the plant,

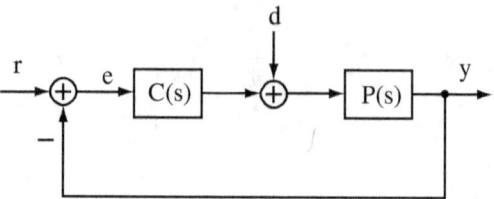

Figure 4.8: Feedback system with disturbance

as shown in Fig. 4.8. The reference input $r(t)$ is now assumed to be zero, but may be added later using superposition (since the systems are linear time-invariant).

For $r(t) = 0$, the Laplace transform gives

$$E(s) = -P(s)\left[D(s) + C(s)E(s)\right], \tag{4.21}$$

so that

$$E(s) = \frac{-P(s)}{1 + P(s)C(s)}D(s). \tag{4.22}$$

For a constant disturbance $d(t) = d_m$, the steady-state error is

$$e_{ss} = \frac{-P(0)}{1 + P(0)C(0)}d_m, \tag{4.23}$$

under the assumption that the closed-loop system is stable.

In terms of the poles and zeros of the plant and compensator

$$\frac{P(s)}{1 + P(s)C(s)} = \frac{n_p(s)d_c(s)}{n_p(s)n_c(s) + d_p(s)d_c(s)}. \tag{4.24}$$

The steady-state error is zero if and only if

$$n_p(0) = 0 \quad \text{or} \quad d_c(0) = 0. \tag{4.25}$$

Note that zero steady-state error is achieved if the plant has a zero at the origin, a fortunate but uninteresting case, since the plant then rejects *all* constant signals, not just disturbances. Therefore, **perfect rejection of constant disturbances** requires that:

- the closed-loop system is stable.

- $C(s)$ has a pole at the $s = 0$.

Disturbance rejection could be obtained for a specific disturbance of known magnitude by subtracting an estimate of the signal d from the control input. However, this would require perfect knowledge of the disturbance, while the result obtained here is achieved despite uncertainties about the disturbance and the plant parameters (as long as stability is preserved).

If perfect tracking *and* perfect disturbance rejection are desired, then the system must be closed-loop stable and the compensator must have a pole at $s = 0$. Because a pole at the origin is associated with an integrator, this strategy is usually referred to as *integral control,* and is extremely common in feedback systems.

The concept can be extended to the tracking/rejection of signals with multiple poles at the origin. For example, perfect tracking of ramp inputs $r(t) = r_m t$ for a constant r_m requires that $C(s)$ must have two poles at $s = 0$. In general, one says that a control system is *of type n* if it has n poles at $s = 0$. Finally, the concept can also be extended to systems with poles on the imaginary axis other than at the origin. For example, perfect tracking of a sinusoidal signal $\sin (\omega_0 t)$ can be achieved by placing compensator poles at $s = \pm j\omega_0$.

4.3.3 Example of integral control

The simplest integral controller is

$$C(s) = \frac{k_I}{s}. \tag{4.26}$$

In the time domain

$$u(t) = k_I \int_0^t e(\tau)d\tau. \tag{4.27}$$

Assume that the plant is a constant gain, or that the plant is stable and that one approximate the plant by its DC gain, *i.e.,*

$$P(s) \simeq P(0). \tag{4.28}$$

Next, let the integral gain be

$$k_I = gP^{-1}(0), \tag{4.29}$$

where $g > 0$ is an adjustable gain. This choice of compensator is shown in Fig. 4.9.

With the approximation (4.28), the response of the system is

$$Y(s) = P(0)D(s) + P(0) \left(\frac{gP^{-1}(0)}{s} (R(s) - Y(s)) \right), \tag{4.30}$$

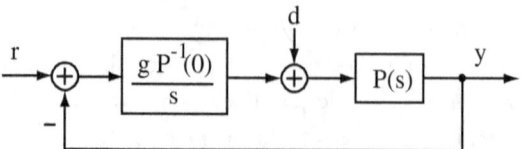

Figure 4.9: Integral control based on the steady-state response of the plant

so that

$$Y(s) = \frac{g}{s+g}R(s) + \frac{sP(0)}{s+g}D(s). \qquad (4.31)$$

The transfer function from the reference input to the output is a stable first-order system with a pole at $s = -g$ and a unity DC gain. The transfer funtion from the disturbance is also a stable first-order system with a pole at $s = -g$ but with a zero DC gain. Therefore, constant reference inputs are perfectly tracked and constant disturbances are perfectly rejected. The speed of response of the system can be directly controlled by choice of the gain g.

Without the steady-state approximation

$$Y(s) = \frac{gP^{-1}(0)P(s)}{s + gP^{-1}(0)P(s)}R(s) + \frac{sP(0)}{s + gP^{-1}(0)P(s)}D(s). \qquad (4.32)$$

The transfer function from the reference input to the output still has unity DC gain and the transfer function from the disturbance still has zero DC gain. Therefore, the properties remain true as long as the closed-loop system is stable. Further, still assuming stability, the results hold despite possible errors in the estimate of $P(0)$.

In general, if the gain g is small, the poles of the closed-loop system will be approximately those of the plant, plus a pole close to $s = -g$. This integral control method is effective in providing tracking control for a stable system, although not necessarily one that provides the fastest possible responses.

In contrast, consider the feedforward approach shown in Fig. 4.10. This system will also provide unity DC gain from the reference input to the output. There is no issue of stability with this controller if the plant is stable. However, there is also no rejection of the disturbance, and any error in $P(0)$ results in an error in the tracking of the reference.

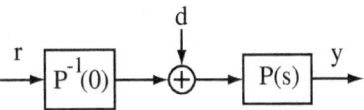

Figure 4.10: Feedforward control based on the steady-state gain of the plant

4.3.4 Proportional-integral-derivative control

An extension of the integral controller is the *proportional-integral-derivative* control law

$$u(t) = k_P(t)e(t) + k_I \int_0^t e(\tau)d\tau + k_D\frac{de(t)}{dt}, \qquad (4.33)$$

where k_P, k_I, k_D are positive, adjustable gains called the *proportional, integral, and derivative gains*, respectively. The transfer function of the control system is

$$C(s) = k_P + \frac{k_I}{s} + k_D s = \frac{k_I + k_P s + k_D s^2}{s}. \qquad (4.34)$$

This type of controller is most common in industry, and is referred to as a *PID controller*. The addition of an integrator in the feedback loop is beneficial for tracking, but detrimental to stability. The proportional and derivative gains give degrees of freedom that, in general, make it possible to maintain stability with faster response times. In the controller, the derivative component is typically filtered, for example by replacing (4.34) by

$$C(s) = k_P + \frac{k_I}{s} + k_D\frac{as}{s+a}, \qquad (4.35)$$

where $a > 0$ is sufficiently large to approximate the derivative.

It is not uncommon to add a feedforward term in the PID controller, as shown in Fig. 4.11. Such controller is called a *two degree-of-freedom* controller and mixes feedforward and feedback control actions. Such structure can be used to fine-tune the properties of the closed-loop transfer function. A *windup* problem also occurs when input limits are encountered and the behavior of the system ceases to be linear. To alleviate this problem, *anti-windup* protection is incorporated into PID control algorithms.

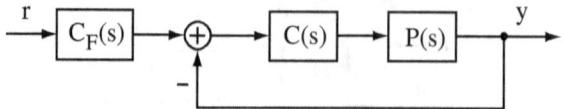

Figure 4.11: Two degree-of-freedom controller mixing feedforward and feedback control

4.4 Effect of initial conditions

The response to nonzero initial conditions was found in (3.81) to be of the form

$$Y(s) = \frac{n_p(s)}{d_p(s)}U(s) + \frac{n_0(s)}{d_p(s)}, \tag{4.36}$$

where $n_p(s)$ and $d_p(s)$ are the numerator and denominator polynomials of the transfer function $P(s)$, and $n_0(s)$ is a polynomial depending on the initial conditions. For the system of Fig. 4.8

$$U(s) = \frac{n_c(s)}{d_c(s)}\left(R(s) - Y(s)\right) + D(s), \tag{4.37}$$

where $n_c(s)$ and $d_c(s)$ are the numerator and denominator polynomials of the transfer function $C(s)$. Combining the two equations

$$\begin{aligned} d_p(s)d_c(s)Y(s) &= n_p(s)n_c(s)R(s) - n_p(s)n_c(s)Y(s) \\ &\quad + d_c(s)n_p(s)D(s) + d_c(s)n_0(s), \end{aligned} \tag{4.38}$$

so that

$$Y(s) = \frac{n_p(s)n_c(s)}{d_{CL}(s)}R(s) + \frac{n_p(s)d_c(s)}{d_{CL}(s)}D(s) + \frac{d_c(s)n_0(s)}{d_{CL}(s)}, \tag{4.39}$$

where

$$d_{CL}(s) = n_p(s)n_c(s) + d_p(s)d_c(s). \tag{4.40}$$

The overall response is the sum of the response to the reference input, the response to the disturbance, and the response to the initial conditions. For all three components, the denominator polynomial is the closed-loop polynomial $d_{CL}(s)$. In all regards, the poles of the system are moved by the feedback. An unstable system can be stabilized, including its response to initial conditions. In contrast, pole/zero cancellation in the feedforward control scheme of Fig. 4.3 would eliminate poles from the response to the reference input, but would not modify the response to the disturbance or the response to the initial conditions.

4.5   Routh-Hurwitz criterion

4.5.1   Background

The Routh-Hurwitz stability test [29] addresses the following question: given a polynomial

$$D(s) = s^n + a_{n-1}s^{n-1} + \cdots + a_0 = 0, \qquad (4.41)$$

find necessary and sufficient conditions such that all the roots of $D(s)$ are in the *open left half-plane* (OLHP), *i.e.*, with $\mathrm{Re}(s) < 0$. Some partial answers to this problem are simple, specifically:

- all roots of $d(s) = s^2 + a_1 s + a_0$ are in the OLHP $\Longleftrightarrow a_1 > 0, a_0 > 0$

- all roots of $d(s) = s^n + a_{n-1}s^{n-1} + \cdots + a_0$ are in the OLHP $\Rightarrow a_{n-1} > 0,$
  $\cdots, a_0 > 0$

Therefore, if any coefficient is zero or negative, there *must* be a root on the $j\omega$-axis or in the open right half-plane. For polynomials of degree 2, the condition is necessary and sufficient, but it is only necessary for higher degrees. In other words, it is possible for a third order polynomial with all positive coefficients to have some roots with $\mathrm{Re}(s) \geqslant 0$. The coefficients must satisfy additional conditions, specified by the Routh-Hurwitz criterion, in order for all the poles to be in the OLHP.

4.5.2   Procedure for the Routh-Hurwitz criterion

The procedure to be followed is:

1. using the coefficients of the polynomial, form an array of numbers (as described below).

2. check the coefficients of the first column. The polynomial has all roots in the open left half-plane $\Longleftrightarrow$ all the coefficients of the first column are nonzero and of the same sign.

Routh array: the construction of the array proceeds as follows. Given

$$D(s) = a_n s^n + a_{n-1}s^{n-1} + \cdots + a_0, \qquad (4.42)$$

1. Create the first two rows using the coefficients of the polynomial and the following pattern

$$
\begin{array}{c|cccc}
s^n & a_n & a_{n-2} & a_{n-4} & a_{n-6} \\
s^{n-1} & a_{n-1} & a_{n-3} & a_{n-5} & \cdots
\end{array}
$$

When a_0 is reached, fill the rest of the array with zeros or leave blank. Label the first row s^n and the second row s^{n-1}.

2. Compute the third row, labelled s^{n-2}, as shown in Fig. 4.12.

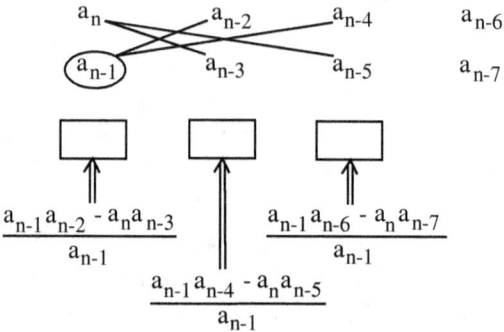

Figure 4.12: Construction of the Routh array

3. Repeat the procedure for additional rows until s^0 is reached.

Comment: if $a_n \neq 1$, the polynomial can be divided by a_n, with no change to the roots. Once $a_n = 1$, the condition to be satisfied is that all the coefficients of the first column of the Routh array must be strictly positive.

4.5.3   Examples

Example 1: consider the polynomial

$$
\begin{aligned}
D(s) &= \left(s^2 + 2s + 5\right)\left(s^2 + 4s + 4\right) \\
&= s^4 + 6s^3 + 17s^2 + 28s + 20. \qquad (4.43)
\end{aligned}
$$

The polynomial has all positive coefficients and has all roots in the open left half-plane (its roots are at $-1 \pm 2j$ and -2 (double root)). The Routh array is given by

s^4	1	17	20	0
s^3	6	28	0	0
s^2	$\frac{6 \times 17 - 28}{6} = \frac{37}{3}$	20	0	0
s^1	$\frac{37/3 \times 28 - 6 \times 20}{37/3} = \frac{676}{37}$	0	0	0
s^0	20	0	0	0

Because the coefficients of the first column are all positive, the test confirms that all the roots are in the open left half-plane.

Example 2: consider the polynomial

$$D(s) = \left(s^2 - 2s + 5\right)\left(s^2 + 4s + 4\right)$$
$$= s^4 + 2s^3 + s^2 + 12s + 20. \tag{4.44}$$

Again, the coefficients are all positive, but we would be mistaken to infer that all the roots are in the open left half-plane. Indeed, the roots may be computed to be $1 \pm 2j$, and -2 (double root). This time, the array is computed to be

s^4	1	1	20	0
s^3	2	12	0	0
s^2	$\frac{2 \times 1 - 1 \times 12}{2} = -5$	$\frac{2 \times 20 - 1 \times 0}{2} = 20$	0	0
s^1	$\frac{-5 \times 12 - 2 \times 20}{-5} = 20$	$\frac{-5 \times 0 - 2 \times 0}{-5} = 0$	0	0
s^0	20			

The coefficients of the first column are not all of the same sign, which confirms the fact that the roots are not all in the open left half-plane.

Example 3: a most interesting feature of the Routh-Hurwitz criterion is that it may be applied to polynomials with variables as coefficients, as opposed to a polynomial with fixed coefficients. In the context of feedback systems, this feature translates into an ability to find conditions on controller parameters that ensure closed-loop stability. Fig. 4.13 shows an example of such an application.

The compensator is a gain k whose value is a free parameter. The closed-loop transfer function is given by

$$P_{CL}(s) = \frac{k/(s+1)^3}{1 + k/(s+1)^3} = \frac{k}{s^3 + 3s^2 + 3s + (1+k)}. \tag{4.45}$$

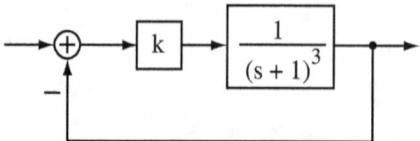

Figure 4.13: Example for the Routh-Hurwitz criterion

The Routh array for this system is

s^3	1	3	0
s^2	3	$(1+k)$	0
s^1	$\dfrac{9-(1+k)}{3}$	0	0
s^0	$1+k$		

and shows that stability is obtained if and only if $1+k>0$ and $8-k>0$, *i.e.,*
if and only if

$$-1 < k < 8 \tag{4.46}$$

This interval is the range of gain k for which the system of Fig. 4.13 is stable.

4.5.4 Explanation of the Routh array

The first two rows of the array contain the coefficients of the polynomials

$$
\begin{aligned}
p_1(s) &= a_n s^n + a_{n-2} s^{n-2} + \dots \\
p_2(s) &= a_{n-1} s^{n-1} + a_{n-3} s^{n-3} + \dots
\end{aligned}
\tag{4.47}
$$

where the elements that are zero by construction are omitted from the array. A
polynomial $p_3(s)$ is defined that is the remainder of the polynomial division of
$p_1(s)$ by $p_2(s)$. Therefore

$$p_1(s) = q_1(s)p_2(s) + p_3(s), \tag{4.48}$$

where $q_1(s) = a_n\, s/a_{n-1}$ is the quotient. The third row of the array contains
the coefficients of the remainder

$$p_3(s) = (a_{n-2} - a_{n-3}\, a_n/a_{n-1})s^{n-2} + (a_{n-4} - a_{n-5}\, a_n/a_{n-1})s^{n-4} + \dots \tag{4.49}$$

Repeating the procedure, polynomials $p_k(s)$ are constructed that are of the form

$$p_k(s) = c_k s^{n-k+1} + \ldots \tag{4.50}$$

where $c_1 = a_n$ and $c_2 = a_{n-1}$. The polynomials alternate as even and odd polynomials of decreasing order. The Routh array contains the coefficients of these polynomials, omitting the coefficients that are always equal to zero due to the even/odd property. The labels on the left of the array give the highest power of s of the polynomials.

Together with the polynomials $p_k(s)$, the procedure also generates a sequence of polynomials $p_k(s) + p_{k+1}(s)$, starting from the original polynomial $p(s) = p_1(s) + p_2(s)$. A key property is that $p_k(s) + p_{k+1}(s)$ has the same number of right half-plane roots and the same number of left half-plane roots as the polynomial $(c_k s + c_{k+1})(p_{k+1}(s) + p_{k+2}(s))$, assuming nonzero leading coefficients up to that point. Remarkably, the imaginary roots of the two polynomials are identical. If there are no zero leading coefficients, the key property implies that the number of right half-plane roots is equal to the number of sign reversals in the first column.

Over the years, approaches were found to simplify the original proof of [29]. A tutorial presentation is available in [4]. Procedures have been developed to extend the solution to cases where a coefficient of the first column is zero. However, the system is known to have poles outside $\text{Re}(s) < 0$ as soon as a sign change or zero coefficient is reached.

4.6   Root-locus method

4.6.1   Motivation

A typical application of the root-locus method is to the standard feedback system of Fig. 4.14, where the compensator is assumed to be of the form $C(s) = kC_0(s)$, i.e., a fixed compensator together with an adjustable gain parameter k. If we let $G(s) = C_0(s)P(s)$, the feedback system takes the form of Fig. 4.15, which is considered in this section.

The root-locus method [10] answers the question of how the poles of the closed-loop system vary as $k = 0 \rightarrow \infty$. Assuming that the open-loop transfer function is written as

$$G(s) = \frac{N(s)}{D(s)}, \tag{4.51}$$

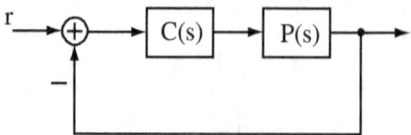

Figure 4.14: Standard feedback system

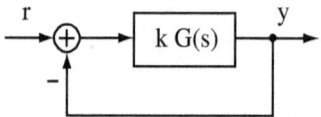

Figure 4.15: Feedback system for the root-locus method

the closed-loop transfer function is given by

$$\frac{Y(s)}{R(s)} = \frac{kG(s)}{1 + kG(s)} = \frac{kN(s)}{D(s) + kN(s)}. \tag{4.52}$$

Therefore, *the root-locus is the locus of the roots of the polynomial* $D(s) + kN(s)$ *for* $k = 0 \rightarrow \infty$. We will consider proper systems ($\deg D(s) \geqslant \deg N(s)$), so that the number of closed-loop poles is equal to the number of open-loop poles.

We begin with some examples to gain insight into what a root-locus may look like.

Example 1: consider the system with

$$G(s) = \frac{1}{s(s+2)} \Rightarrow D(s) + kN(s) = s^2 + 2s + k. \tag{4.53}$$

The roots are given by

$$s^2 + 2s + k = 0 \Rightarrow s = \frac{-2 \pm \sqrt{4 - 4k}}{2} = -1 \pm \sqrt{1 - k}. \tag{4.54}$$

For a few values of k, we have the closed-loop poles

k	s_1	s_2
0	-2	0
0.5	$-1 + \sqrt{0.5} = -1.7$	$-1 - \sqrt{0.5} = -0.3$
1	-1	-1
2	$-1 + j$	$-1 - j$
5	$-1 + 2j$	$-1 - 2j$
101	$-1 \pm 10j$	$-1 \pm 10j$

From these results, the locus of the two poles as k varies from 0 to ∞ can be deduced to be as shown in Fig. 4.16. In general, the root-locus is described by smooth curves, or branches, whose number is equal to the number of open-loop poles. Note that, for this example, the response of the system is stable for all k, but becomes oscillatory for $k > 2$.

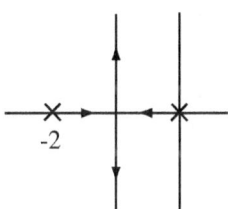

Figure 4.16: Root-locus for example 1

Example 2: add a zero to the system, so that

$$G(s) = \frac{s + 1}{s(s + 2)} \Rightarrow D(s) + kN(s) = s^2 + (2 + k)s + k. \tag{4.55}$$

In this case, the closed-loop poles are given by

$$s_{1,2} = \frac{-(2 + k) \pm \sqrt{(2 + k)^2 - 4k}}{2} = \frac{-(2 + k) \pm \sqrt{4 + k^2}}{2}. \tag{4.56}$$

The two poles are real for all k. A representative set of values for the poles is given below.

k	s_1	s_2
0	0	-2
1	-0.4	-2.6
100	-0.99	-101
∞	-1	k

and the root-locus may be deduced to be the one shown in Fig. 4.17.

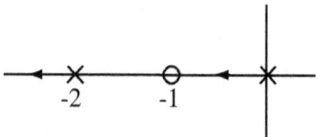

Figure 4.17: Root-locus for example 2

Example 3: the next example is similar to the previous one, but with the location of the nonzero pole and the zero reversed. Specifically

$$G(s) = \frac{s+2}{s(s+1)} \Rightarrow D(s) + kN(s) = s^2 + s(1+k) + 2k, \qquad (4.57)$$

and the poles are given by

$$s_{1,2} = \frac{-(1+k) \pm \sqrt{(1+k)^2 - 8k}}{2}. \qquad (4.58)$$

Whether the poles are real or complex is determined by the sign of the function

$$f(k) = (1+k)^2 - 8k = k^2 - 6k + 1, \qquad (4.59)$$

which has roots at 0.2 and 5.8. Therefore, the function $f(k)$ has the shape shown in Fig. 4.18.

In terms of the original polynomial, we may conclude that

$k = 0 \rightarrow 0.2$	2 real roots
$k = 0.2 \rightarrow 5.8$	2 complex roots
$k = 5.8 \rightarrow \infty$	2 real roots

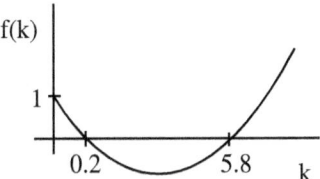

Figure 4.18: Function f(k)

A few representative values of the closed-loop poles may be computed to be

k	s_1	s_2
0	0	−1
⋮	real	real
0.2	−0.6	−0.6
⋮	complex	complex
5.8	−3.4	−3.4
⋮	real	real
100	−2.02	−98.98

and the root-locus may be deduced to be as shown on Fig. 4.19.

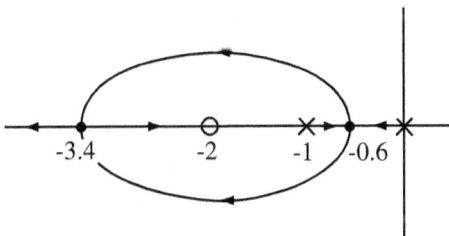

Figure 4.19: Root-locus for example 3

Although the closed-loop poles cannot be computed exactly in general, as was the case for the three examples presented here for motivation, it turns out that the root-locus satisfies enough properties that its general shape can be predicted rather well using simple rules.

4.6.2 Main root-locus rules

Definitions

The open-loop transfer function is assumed to be expressed as

$$P(s)C(s) = k\ G(s) = k\frac{(s - z_1)\,(s - z_2)\ \cdots\ (s - z_m)}{(s - p_1)(s - p_2)\ \cdots\ (s - p_n)}. \tag{4.60}$$

There are m open-loop zeros $z_1, \cdots, z_m$, and n open-loop poles $p_1, \cdots, p_n$. Assume that $m \leqslant n$, and that the numerator and denominator polynomials have real coefficients. The *root-locus* is the locus of the closed-loop poles as $k = 0 \to \infty$. These are the poles of $kG(s)/(1 + kG(s))$ and, therefore, the roots of

$$(s - p_1)(s - p_2)\cdots(s - p_n) + k\,(s - z_1)\,(s - z_2)\cdots(s - z_m) = 0. \tag{4.61}$$

Main root-locus rules

Branches of the root-locus: there are n branches in the root-locus (*i.e.*, n closed-loop poles for all k). The n branches start at the open-loop poles (start means $k = 0$) and m of the branches finish at the open-loop zeros (finish means $k \to \infty$). The root-locus is symmetric with respect to the real axis.

Portions of the real axis: to determine if a portion of the real axis belongs to the locus, count the total number of open-loop poles and open-loop zeros that lie to the right of a point on the real axis. The point belongs to the locus if and only if the number is odd. As a consequence, the portion of the real axis that is farther to the right than any pole or zero does not belong to the locus.

Asymptotes: the $n - m$ poles that do not go to open-loop zeros go to infinity. For k large, the branches become close to straight lines, called *asymptotes*, which:
(a) all intersect at the same point called *centroid*, which is located on the real axis at:

$$\sigma = \frac{\displaystyle\sum_{i=1}^{n} p_i - \sum_{j=1}^{m} z_j}{n - m}. \tag{4.62}$$

(b) form angles with the real axis equal to $\dfrac{180°}{n - m} + i\dfrac{360°}{n - m}$ with $i = 0, 1, \cdots$, $n - m - 1$.

Angles of departure and arrival on the real axis: for an open-loop pole on the real axis with multiplicity r, the angles of departure of the branches are either $i\frac{360°}{r}$ or $\frac{180°}{r} + i\frac{360°}{r}$ (with $i = 0, 1, \cdots, r - 1$). Which case applies can be determined using the rule regarding the portions of the real axis. Branches reach

multiple zeros on the real axis with the same set of angles. When roots merge
on the real axis, one set of angles defines the angles formed by the incoming
branches, and the other set defines the angles formed by the outgoing branches.

Examples

Fig. 4.20 shows examples of application of the basic rules. The angles of the
asymptotes are: $180°$ if $n - m = 1$, $(90°, -90°)$ if $n - m = 2$, $(180°, 60°, -60°)$ if
$n - m = 3$, $(45°, -45°, 135°, -135°)$ if $n - m = 4$,... In cases 1 and 3 of Fig. 4.20,
the angles are $\pm 90°$. In cases 2 and 4, the angles are $\pm 60°$ and $180°$. In general,
application of the rules is fairly straightforward, although inferring the shape
of the root-locus becomes easier with experience, and sometimes requires some
amount of guessing.

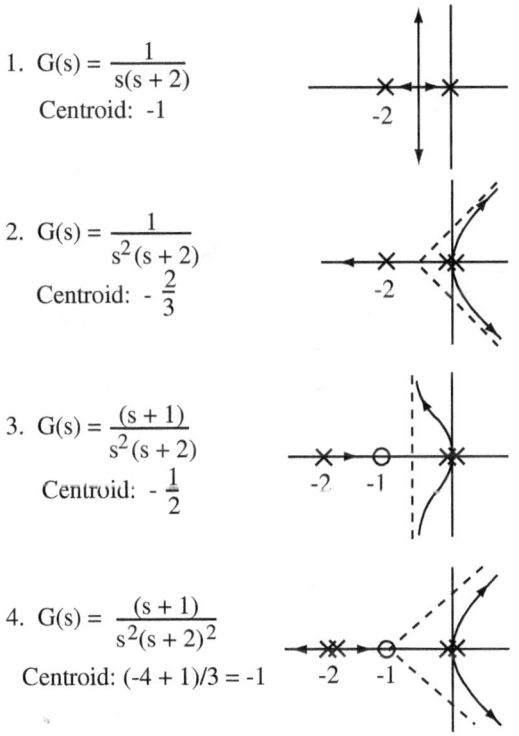

1. $G(s) = \dfrac{1}{s(s + 2)}$

   Centroid: -1

2. $G(s) = \dfrac{1}{s^2(s + 2)}$

   Centroid: $-\dfrac{2}{3}$

3. $G(s) = \dfrac{(s + 1)}{s^2(s + 2)}$

   Centroid: $-\dfrac{1}{2}$

4. $G(s) = \dfrac{(s + 1)}{s^2(s + 2)^2}$

   Centroid: $(-4 + 1)/3 = -1$

Figure 4.20: Root-locus examples

For the angles of departure on the real axis, the set of possible angles are
shown on Fig. 4.21. In other words, the possible patterns are the same as for
the asymptotes, plus the patterns rotated by $180°$ divided by the number of

poles. In cases 2, 3 and 4 of Fig. 4.20, poles leave at $\pm 90°$ (the other pattern is excluded because the portions of the real axis on both sides of the poles do not belong to the root-locus). The same set of angles applies when poles merge together on the real axis. Then, poles reach the so-called *breakaway point* with one set of angles, and they leave with the other. In case 1 of Fig. 4.20, poles merge with incoming branches at $(0°, 180°)$ and with outgoing branches at $(90°, -90°)$. Note that, although not shown on the examples, the same set of angles also applies when branches reach multiple zeros on the real axis.

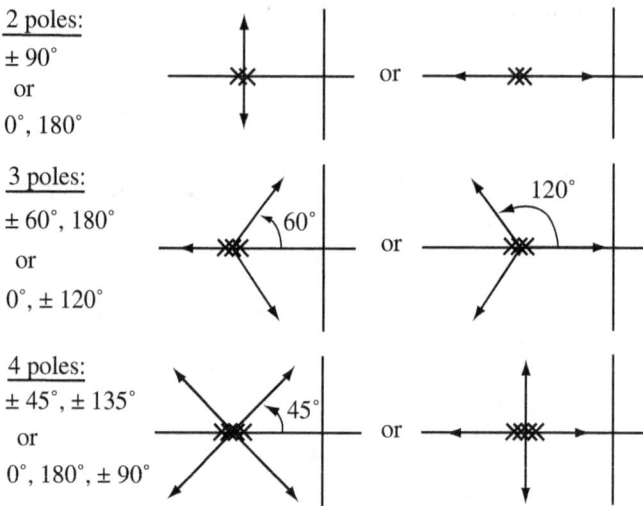

2 poles:
$\pm 90°$
or
$0°, 180°$

3 poles:
$\pm 60°, 180°$
or
$0°, \pm 120°$

4 poles:
$\pm 45°, \pm 135°$
or
$0°, 180°, \pm 90°$

Figure 4.21: Angles of departure from the real axis

4.6.3 Additional root-locus rules

Often, the most important characteristics of the root-locus can be deduced using only the basic rules. In this manner, important information about the poles of a closed-loop system can be obtained remarkably quickly. Sometimes, however, additional rules prove useful to refine the root-locus. The rules are presented first, followed by examples illustrating their application.

Additional root-locus rules

breakaway points from the real axis: points where the root-locus breaks away from the real axis are the real roots of the equation $\frac{d}{ds}[G(s)] = 0$. The reverse is not necessarily true: there may be complex solutions to the equation, and some solutions may correspond to $k < 0$. One should evaluate $k = -D(s)/N(s)$ at the roots. If k is real and $k > 0$ for some root, the root gives the location of a breakaway point from the real axis.

Crossing of the $j\omega$-axis: the values of k for which the root-locus crosses the $j\omega$-axis can be determined by applying the Routh-Hurwitz criterion to (4.61). Given k, the roots of (4.61) determine the locations where the branches cross the $j\omega$-axis.

Angles of departure and arrival: the angle θ between the tangent to the root-locus close to an open-loop pole and the direction of the real axis can be determined as follows. Assume that the angle of departure is calculated for the pole p_1 (the procedure can be repeated for other poles in a similar manner). Assume that the pole is not repeated. Let α_i be the angle between the direction of the real axis and the vector drawn from the pole p_i to the pole p_1. Let β_j be the similar angles for the zeros. The angle θ is given by

$$\theta = 180° - \sum_{i=2}^{n} \alpha_i + \sum_{j=1}^{m} \beta_j. \tag{4.63}$$

For a repeated pole with multiplicity r, (4.63) is replaced by

$$\theta = \frac{1}{r}\left(180° - \sum_{i=r+1}^{n} \alpha_i + \sum_{j=1}^{m} \beta_j + l\,360°\right) \qquad l = 0, \cdots, r-1. \tag{4.64}$$

The procedure can also be applied to determine the angles of arrival to the zeros. In this case, the procedure is identical, except that θ is replaced by $-\theta$. Therefore, for the angle of arrival to a zero of multiplicity r

$$\theta = \frac{-1}{r}\left(180° - \sum_{i=1}^{n} \alpha_i + \sum_{j=r+1}^{m} \beta_j + l\,360°\right) \qquad l = 0, \cdots, r-1 \tag{4.65}$$

Example - breakaway points from the real axis: consider the system

$$G(s) = \frac{(s+2)}{s(s+1)}, \tag{4.66}$$

whose root-locus was obtained before and is shown in Fig. 4.22. The breakaway points had already been determined to be at -0.6 and -3.4. We may verify these values using the fact that

$$\frac{dG(s)}{ds} = \frac{s(s+1) - (2s+1)(s+2)}{s^2(s+1)^2}, \tag{4.67}$$

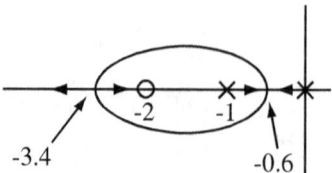

Figure 4.22: Example of breakaway points

so that

$$\frac{dG(s)}{ds} = 0 \iff s^2 + 4s + 2 = 0. \tag{4.68}$$

The roots of this polynomial are $s_{1,2} = -2 \pm \sqrt{2} = -0.6$ and -3.4, which confirms the earlier result. In general, however, one should verify that the root is really a breakaway point by computing k. For $s = -2 + \sqrt{2} = -0.6$,

$$k = -\frac{D(s)}{N(s)} = -\frac{\left(-2 + \sqrt{2}\right)\left(-1 + \sqrt{2}\right)}{\sqrt{2}} = 0.17. \tag{4.69}$$

Since k is real and $k > 0$, the breakaway point belongs to the root-locus. The same property can be checked for the other root.

Example - crossing of the $j\omega$−axis: consider the system

$$G(s) = \frac{1}{(s+1)^3} \tag{4.70}$$

In this case, the closed-loop poles may be computed exactly, and are given by

$$(s + 1) = ^3\sqrt{-k} \Rightarrow s = -1 + \sqrt[3]{k}\,\sqrt[3]{-1}, \tag{4.71}$$

or

$$\begin{aligned}
s_1 &= -1 + \sqrt[3]{k}\; e^{j\pi/3}, \\
s_2 &= -1 + \sqrt[3]{k}\; e^{j2\pi/3}, \\
s_3 &= -1 + \sqrt[3]{k}\; e^{-j2\pi/3}.
\end{aligned} \tag{4.72}$$

The root-locus is shown in Fig. 4.23, and may be shown to satisfy the root-locus rules. We may easily determine that crossing of the imaginary axis occurs when

$$\sqrt[3]{k}\; \cos(60°) = 1, \text{ or } k = 8. \tag{4.73}$$

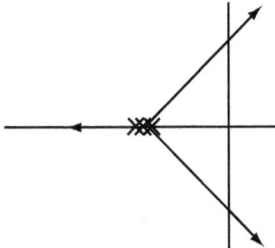

Figure 4.23: Example of crossing of $j\omega$-axis

In general, the roots cannot be computed analytically, and one may apply the Routh-Hurwitz criterion to find the range. This computation was done before and yielded (4.46). The locations of the crossings of the imaginary axis are obtained by letting $k = 8$ in the closed-loop polynomial. With $s^3 + 3s^2 + 3s + 9 = 0$, the roots are $s = -3$, $s = \pm j\sqrt{3}$, and the complex roots of this equation correspond to the crossings.

Example - angles of departure: consider the system

$$G(s) = \frac{s+3}{(s^2 + 2s + 2)(s + 2)}.$$ (4.74)

One denotes

$$\theta = \text{the angle of departure from the given pole } p_1$$
$$\alpha_i = \text{the angle from the } i^{th} \text{ pole to the given pole } p_1$$
$$\beta_j = \text{the angle from the } j^{th} \text{ zero to the given pole } p_1$$

Let the pole $p_1 = -1 + j$. The angles are shown on Fig. 4.24.

The rule says that

$$\theta = 180° - \alpha_2 - \alpha_3 + \beta_1.$$ (4.75)

From the figure and with $\tan^{-1}(0.5) = 26.6°$, the angle of departure from the pole is

$$\theta = 180° - 90° - 45° + \tan^{-1}(0.5) = 71.6°.$$ (4.76)

The result is consistent with the figure as it was drawn.

As a second example, consider the system

$$G(s) = \frac{s+1}{(s+2)(s^2+1)}.$$ (4.77)

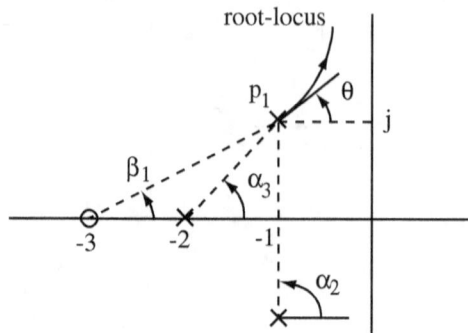

Figure 4.24: Definition of angles for the computation of the angle of departure

A possible root-locus is shown on the left of Fig. 4.25. However, one may be concerned that the branches leaving the complex poles could cross into to the right half-plane as shown on the right of Fig. 4.25, making the system unstable for small gain..

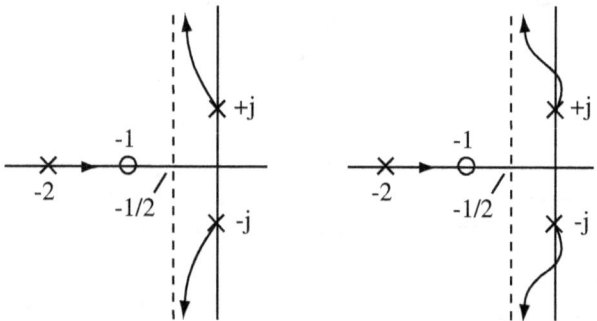

Figure 4.25: Two possibilities for the root-locus of the second example for angles of departure

The relevant angles for the computation of the angle of departure from the pole at $s = j$ are shown in Fig. 4.26. Since $\tan^{-1}(0.5) = 26.6°$, the formula gives

$$\theta = 180° - 90° - 26.6° + 45° = 108.4°. \tag{4.78}$$

Therefore, the branch leaving $s = j$ indeed leaves the pole as shown on Fig. 4.26 and on the left of Fig. 4.25.

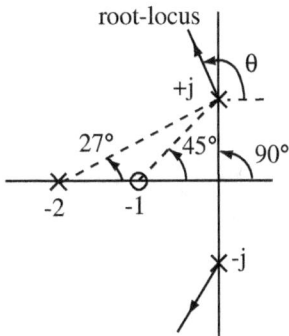

Figure 4.26: Second example for angles of departure

Example - angles of arrival: consider the system

$$G(s) = \frac{s^2 + 4}{s(s+1)^2}.$$

(4.79)

Using the main rules, the root-locus for this system may be drawn as shown in Fig. 4.27. The figure also gives the angles for the computation of the angle of arrival at $s = 2j$.

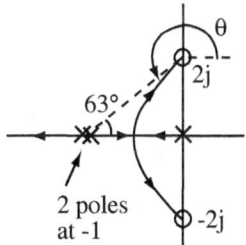

Figure 4.27: Example for angle of arrival

The formula gives

$$-\theta = 180° \underbrace{-63° - 63°}_{\text{2 poles at -1}} - \underbrace{90°}_{\text{pole at 0}} + \underbrace{90°}_{\text{zero at } 2j} \quad \Rightarrow \theta = -54°.$$

(4.80)

Oddly, this result forces us to redraw the root-locus of Fig. 4.27 to become the one shown in Fig. 4.28. In fact, the system becomes unstable for large gain, which would not have been predicted from the tentative Fig. 4.27.

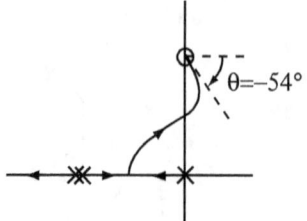

Figure 4.28: Example for angle of arrival with actual shape of the root-locus

Example - angles of departure/arrival for multiple poles/zeros: for
multiple poles or zeros, one uses the same formula and divides the result by the
multiplicity r. Adding multiples of $360°/r$ gives the other angles. Consider

$$G(s) = \frac{(s+1)}{(s^2+1)^2},\tag{4.81}$$

whose root-locus is shown in Fig. 4.29. The angles of departure for the two poles
at $s = j$ are given by

$$2\theta = 180° \underbrace{-2 \times 90°}_{\text{poles at } -j} + \underbrace{45°}_{\text{zero at } -1} + \left\{ \begin{array}{c} 0° \\ 360° \end{array} \right\} \Rightarrow \theta = 22.5°,\ 202.5°.\tag{4.82}$$

These values are consistent with the root-locus as drawn on the figure.

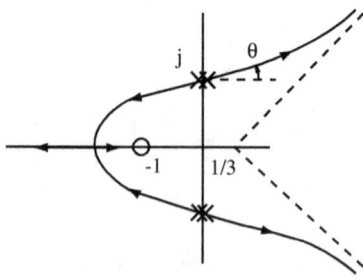

Figure 4.29: Example of angle of departure for multiple poles

Note that this example is a good opportunity to compute the location of the
breakaway points on the real axis. Specifically,

$$\frac{dG(s)}{ds} = \frac{(s^2+1)^2 - 2(s^2+1)2s(s+1)}{(s^2+1)^2} = 0\tag{4.83}$$

if and only if

$$\left(s^2 + 1\right) - 4s(s + 1) = -3s^2 - 4s + 1 = 0, \tag{4.84}$$

whose roots are at $s_1 = -1.55$ and $s_2 = 0.22$. The values of gains k corresponding to the two roots are given by

$$k_1 = -1/G(s_1) = 21, \quad k_2 = -1/G(s_2) = -0.9. \tag{4.85}$$

The second root turns out to be a breakaway point for the complementary root-locus, or root-locus for $k < 0$, which is discussed next.

4.6.4   Complementary root-locus

The root-locus for $k = 0 \rightarrow -\infty$ is called the *complementary root-locus*. Since Fig. 4.15 assumed negative feedback with $k > 0$, the case $k < 0$ corresponds to positive feedback. The rules for $k < 0$ are similar to those for the regular root-locus. The most significant differences are that the portions of the real axis are those that do not belong to the regular root-locus and that the angles of the asymptotes are the alternate patterns shown in Fig. 4.21.

Complementary root-locus rules

The rules are the same as for $k > 0$, except that:

Branches of the root-locus: when $m = n$, one or more branches of the root-locus may reach ∞ for $k = -1$.

Portions of the real axis: replace "odd" by "even". The portion of the real axis to the right of the rightmost pole or zero now belongs to the locus.

Asymptotes: replace the angles by $i\frac{360°}{n - m}$.

Angles of departure: replace $180°$ by $0°$.

Examples

Generally, the angles of the asymptotes are now $0°$ if $n - m = 1$, $(0°, 180°)$ if $n - m = 2$, $(0°, 120°, -120°)$ if $n - m = 3$ and $(0°, 180°, 90°, -90°)$ if $n - m = 4$.

Example 1: Fig. 4.30 shows the example of

$$G(s) = \frac{s + 1}{s^2(s + 2)^2}, \tag{4.86}$$

which corresponds to $n - m = 3$. The complementary root-locus can be viewed as the second piece of the overall root-locus, which covers the range $k = -\infty \rightarrow \infty$.

Example 2: Another example corresponds to the system considered earlier for the angles of departure from multiple poles, that is

$$G(s) = \frac{(s + 1)}{(s^2 + 1)^2}. \tag{4.87}$$

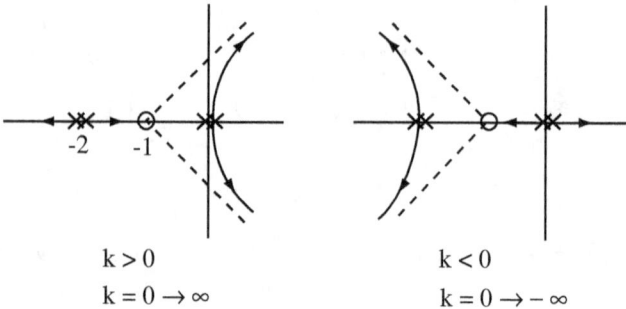

Figure 4.30: Comparison of root-locus and complementary root-locus

Application of the rules yields opposite portions of the real axis, identical centroid, asymptotes at rotated by $180°/(n-m)$ (that is, with angles $0°$ and $\pm120°$),
and a complementary breakaway point at $s = 0.22$. The angles of departure are
also rotated by $180°/r$, that is $90°$ for the two imaginary poles (yielding $112.5°$
and $292.5°$). The resulting complementary root-locus is shown in Fig. 4.31.

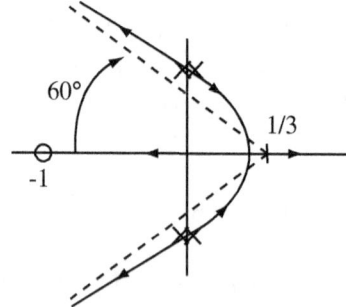

Figure 4.31: Example of complementary root-locus

4.6.5   Important conclusions from the root-locus rules

The main root-locus rules allow one to reach the following useful observations:

1. zeros in the right half-plane always lead to instability for high gain. Such
   zeros are usually undesirable in either the plant or the controller.

2. systems whose number of poles exceed the number of zeros by 3 or more always become unstable for sufficiently high gain. The higher the pole/zero excess, the higher the danger of instability at high gain. Typical control systems have as many zeros as poles.

3. if an open-loop zero is close to an open-loop pole, the pole will move towards that zero. This property can be used in control design to attract poles to desirable locations. An example is shown in Fig. 4.32, assuming that additional poles are somewhat distant from the two main poles. The damping of oscillatory poles can be increased by attracting them to well-placed zeros.

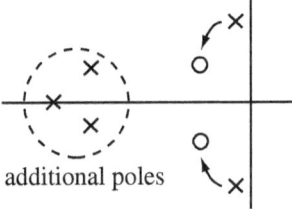

Figure 4.32: Attraction of poles by zeros

4. moving a zero far in the left half-plane may be counterproductive, because the centroid will be pushed towards the right half-plane. Fig. 4.33 gives an example where it is preferable to place the zero closer to the origin.

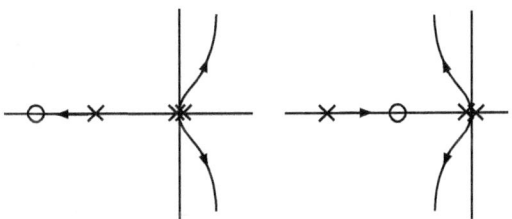

Figure 4.33: Effect of a zero on asymptotes

5. a pole at $s = 0$ in $C(s)$ desirable for zero steady-state error, but tends to make the closed-loop system less stable. For example, Fig. 4.34 shows

options for the control of the system with transfer function

$$P(s) = \frac{k}{s(s+a)}. \tag{4.88}$$

Four controllers are considered: (a) proportional, (b) proportional-derivative,

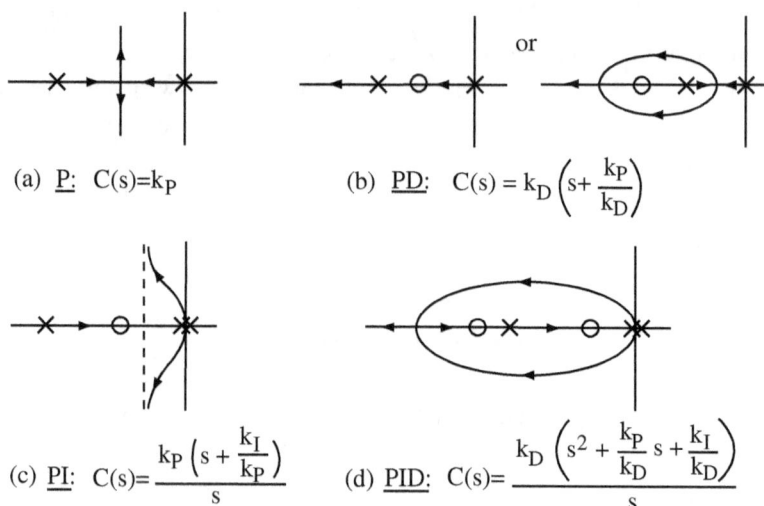

(a) <u>P</u>: $C(s) = k_P$

(b) <u>PD</u>: $C(s) = k_D\left(s + \dfrac{k_P}{k_D}\right)$

(c) <u>PI</u>: $C(s) = \dfrac{k_P\left(s + \dfrac{k_I}{k_P}\right)}{s}$

(d) <u>PID</u>: $C(s) = \dfrac{k_D\left(s^2 + \dfrac{k_P}{k_D}s + \dfrac{k_I}{k_D}\right)}{s}$

Figure 4.34: Choices of compensators for $P(s) = k/(s(s+a))$

(c) proportional-integral, (d) proportional-integral-derivative. Generally, addition of the derivative term improves the damping of the system, while inclusion of integral action makes it more difficult to achieve a satisfactory transient response. From (a) to (c), for example, the asymptote moves towards the right, and the poles stay closer to the imaginary axis for small gain.

6. unmodelled dynamics (additional poles in the actual plant transfer function) may produce instabilities for large gain. For example, a system with transfer function

$$G(s) = \frac{1}{s(s+1)}. \tag{4.89}$$

is stable for all gain, as shown by the root-locus on the left of Fig. 4.35. However, if an additional real pole is present in the actual system, the root-locus becomes the one shown on the right of Fig. 4.35. No matter how

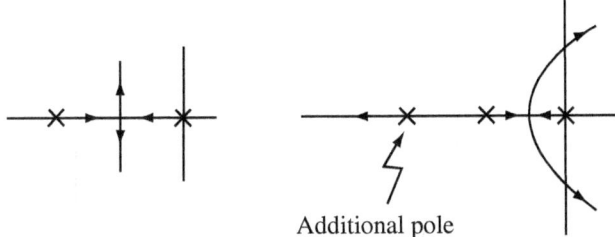

Additional pole

Figure 4.35: Destabilizing effect of an additional pole on a root-locus with two real poles

far the pole is in the left half-plane, the closed-loop system will become unstable for large enough gain.

7. Because the positive real axis belongs to the root-locus for $k < 0$, a system always becomes unstable for positive feedback of large gain. For small gain, however, positive feedback can be stabilizing. For example, consider the root-locus for

$$G(s) = \frac{s + 2}{(s + 1)(s^2 + 1)}. \tag{4.90}$$

The regular root-locus ($k > 0$) is shown on the left of Fig. 4.36, while the complementary root-locus is shown on the right of the figure. For $k > 0$, the two poles at $s = \pm j$ immediately move to the right half-plane, towards the $\pm 90°$ asymptotes with centroid at $s = 1/2$. The angle of departure of the pole at $s = j$ is equal to $\theta = 180° - 90° - 45° + \tan^{-1}(0.5) = 71.6°$. The system is unstable for all gain $k > 0$. In contrast, for $k < 0$, the angle of departure of the pole at $s = j$ becomes $251.6°$ and the pole moves towards the left half-plane. The pole at $s = 1$ moves towards the right half-plane and, for sufficiently large gain, the system becomes unstable. However, for some range of gain, the system is stabilized with $k < 0$.

Root-locus rules are useful to quickly sketch a root-locus and understand how pole and zero locations affect its shape. For complicated cases or for precision plots, a modern software package should be used. For example, the shape of the complementary root-locus of Fig. 4.36 was obtained using the Matlab commands:

```
num=[1 2];
den=conv([1 1],[1 0 1]);
rlocus(-num,den)
```

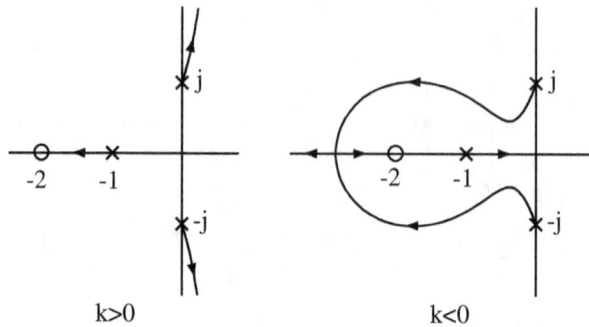

Figure 4.36: Root-locus with positive and negative feedback

The regular root-locus can be obtained by replacing the last line of the code by rlocus(num,den). The *conv* function is convenient to multiply the two denominator polynomials.

4.7 Feedback design for phase-locked loops

4.7.1 Modulation of signals in communication systems

The transmission of signals in telecommunication systems is typically performed through a transmitter and a receiver, as shown in Fig. 4.37. The purpose of the modulator is to shift the frequency spectrum of the signal $x(t)$ to a higher frequency range, so that transmission over electromagnetic waves can be performed efficiently. A demodulator is needed to perform the reverse operation. Ideally, the demodulated signal $x_d(t)$ is equal, or proportional to $x(t)$.

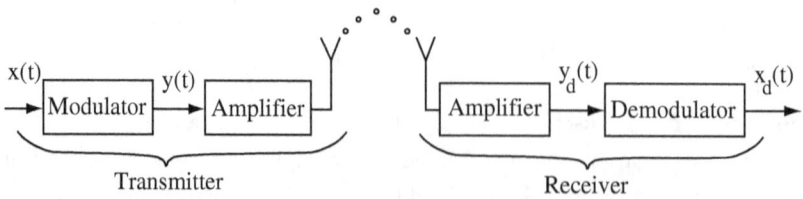

Figure 4.37: Communication system

Two basic methods of modulation are *amplitude modulation* (AM) and *fre-*

quency modulation (FM). In amplitude modulation, one has that

$$y(t) = (A + k_m \, x(t)) \sin (2\pi f_c \, t) . \tag{4.91}$$

Amplitude modulation of a sinusoidal signal $x(t)$ is shown in Fig. 4.38. The frequency f_c is called the *carrier frequency*. In commercial AM radio, f_c ranges from 530 to 1600 kHz. The spectrum of $x(t)$ is limited to 5 kHz. Low-pass filtering is used to limit the spectrum of $x(t)$ to this range.

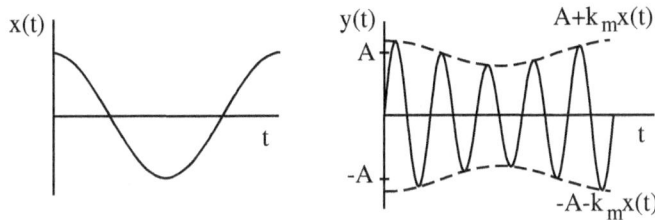

Figure 4.38: Amplitude modulation

In *frequency modulation*, one has

$$
\begin{aligned}
y(t) &= A \sin (\theta(t)), \\
\theta(t) &= 2\pi f_c \, t + 2\pi k_m \int_0^t x(\sigma) d\sigma.
\end{aligned} \tag{4.92}
$$

The frequency f_c is called the *center frequency*. It is the frequency of the signal $y(t)$ when $x(t) = 0$. The *instantaneous frequency* (in Hz) of the signal $y(t)$ is defined to be

$$f(t) = \frac{1}{2\pi} \frac{d\theta(t)}{dt} = f_c + k_m x(t). \tag{4.93}$$

In an implementation with analog electronics, k_m has the units of Hz/V, assuming that $x(t)$ has the units of volts. The parameter k_m specifies how much the frequency of the signal y increases per unit increase of the magnitude of the signal x. Frequency modulation by a sinusoidal signal $x(t)$ is shown in Fig. 4.39.

In commercial FM radio, f_c ranges from 88 to 108 MHz. The spectrum of $x(t)$ is limited to $f_{max} = 53$ kHz, which includes two (stereo) channels with 15 kHz bandwidth each. If the modulating signal is proportional to $x(t)$, instead of the integral of $x(t)$, the modulation is referred to as *angle modulation* or *phase modulation*. In some cases, the modulating signal is *digital* (on/off), which leads

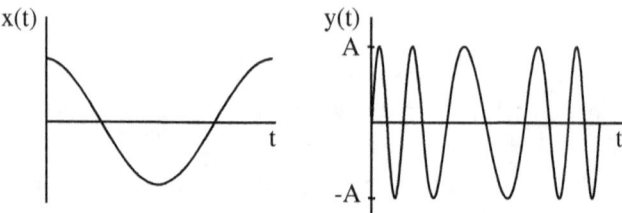

Figure 4.39: Frequency modulation

to *frequency shift keying* (FSK, if $x(t)$ is digital) or *phase shift keying* (PSK, if $\theta(t)$ is digital). For 180° phase reversals, PSK becomes *phase reversal keying* or *binary phase shift keying* (BPSK). For four values of the phase ($\theta = 0°$, 90°, 180°, 270°), one has *quaternary* or *quadriphase* PSK (QPSK).

4.7.2   Voltage-controlled oscillators

A frequency modulator is also called a *voltage-controlled oscillator* (VCO). Such device performs the transformation from a signal $x(t)$ to a signal $y(t)$ whose frequency is determined by the magnitude of $x(t)$. A VCO is also called a *voltage-to-frequency converter* ($V \rightarrow f$ converter). Mathematically, the representation of a VCO is that of (4.92) or Fig. 4.40: it is an integrator followed by a sinusoidal function. Note that the signal $2\pi f_c\, t$ can be moved to the input of the integrator, as a *constant* signal f_c/k_m added to $x(t)$.

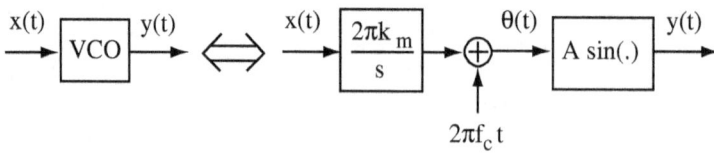

Figure 4.40: Voltage-controlled oscillator and mathematical equivalent

4.7.3   Phase-locked loops

The device that performs the inverse operation of a VCO is called an *FM demodulator* or $f \rightarrow V$ *converter*. A *phase-locked loop* (PLL) is such a device, whose general structure is shown in Fig. 4.41. Interestingly, a PLL includes a

VCO as one of its components. The signal $x_{vco}(t)$ is, under ideal conditions, proportional to the signal that generated $y(t)$, that is $x(t)$. The signal $y_{vco}(t)$ then has the same instantaneous frequency as $y(t)$.

The phase detector is a device that generates a signal $\phi(t)$ whose value is proportional to the difference of phase between the signals $y(t)$ and $y_{vco}(t)$. Much of the complexity of phase-locked loops is related to this device. A filter is typically needed after the phase detector, because of harmonic components associated with practical detectors, and to improve the stability properties of the feedback system.

The concept of a phase-locked loop is that, if there is a small phase error so that $y_{vco}(t)$ lags behind $y(t)$, a signal $\phi(t)$ appears at the output of the detector. This signal produces an increase in the voltage applied to the VCO, so that the frequency of $y_{vco}(t)$ increases and the phase of the signal catches up with the phase of the incoming signal $y(t)$. In steady-state, the phases of $y(t)$ and $y_{vco}(t)$ are equal, or separated by a constant, and the signals are said to be locked in phase.

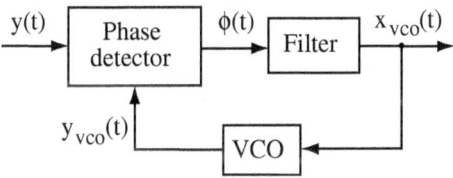

Figure 4.41: Phase-locked loop

Assuming that an ideal phase detector is available, the equations for the system are:

$$\text{VCO (modulator):} \quad y(t) = A \sin(\theta(t)),$$

$$\theta(t) = 2\pi f_c t + 2\pi k_m \int_0^t x(\sigma)d\sigma. \qquad (4.94)$$

$$\text{VCO (PLL):} \quad y_{vco}(t) = A_{vco} \sin(\theta_{vco}(t)),$$

$$\theta_{vco}(t) = 2\pi f_{c,vco} t + 2\pi k_{vco} \int_0^t x_{vco}(\sigma)d\sigma. \qquad (4.95)$$

$$\text{Phase detector:} \quad \phi(t) = k_{pd}(\theta(t) - \theta_{vco}(t)). \qquad (4.96)$$

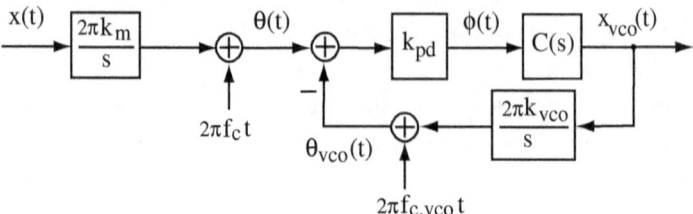

Figure 4.42: Diagram of a phase-locked loop with ideal phase detector

$$\text{Filter:} \qquad X_{vco}(s) = C(s)\,\Phi(s). \qquad (4.97)$$

The gain of the phase detector is k_{pd} and has the units of Volts/rad in an analog phase-locked loop. $C(s)$ is the transfer function of the filter. The instantaneous frequency of the VCO of the PLL is

$$f_{vco}(t) = \frac{1}{2\pi}\frac{d\theta_{vco}(t)}{dt} = f_{c,vco} + k_{vco}x_{vco}(t). \qquad (4.98)$$

The dynamics of the system are highly nonlinear, because of the sinusoidal functions. However, under the assumption of an ideal phase detector, the equations describing the variables $\theta(t)$, $\theta_{vco}(t)$, $x(t)$, and $x_{vco}(t)$ are linear, as shown in Fig. 4.42. The system can therefore be analyzed using linear time-invariant methods (in particular, transfer functions can be used).

Fig. 4.42 can be transformed into a simpler, equivalent diagram, using the following definitions. Denote the difference between the center frequencies of the VCO of the modulator and of the VCO of the demodulator

$$\delta f_c = f_c - f_{c,vco} \qquad (4.99)$$

and let

$$x_s(t) = \frac{k_m}{k_{vco}}x(t). \qquad (4.100)$$

Note that $x_s(t)$ is proportional to $x(t)$. We will find that the signal $x_{vco}(t)$ converges to this scaled signal under ideal conditions.

We have that

$$
\begin{aligned}
\theta(t) - \theta_{vco}(t) &= 2\pi\delta f_c t + 2\pi k_{vco}\int_0^t \left(x_s(\sigma) - x_{vco}(\sigma)\right)d\sigma \\
&= 2\pi k_{vco}\int_0^t \left(x_s(\sigma) - x_{vco}(\sigma) + \frac{\delta f_c}{k_{vco}}\right)d\sigma. \qquad (4.101)
\end{aligned}
$$

Therefore, the diagram of Fig. 4.43 represents the phase-locked loop if one defines the two constants

$$\begin{aligned} k_{pll} &= 2\pi k_{vco} k_{pd}, \\ d_0 &= \frac{\delta f_c}{k_{vco}}. \end{aligned} \qquad (4.102)$$

as well as the disturbance signal $d(t) = d_0$. k_{pll} is the gain of the phase-locked loop, not including the filter $C(s)$.

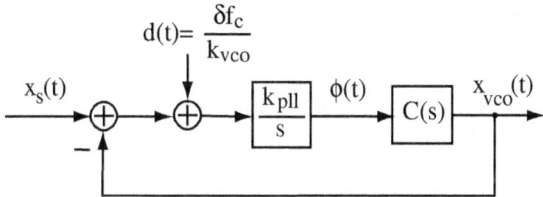

Figure 4.43: Equivalent diagram of a phase locked-loop with ideal phase detector

The diagram of the phase locked-loop is similar to a conventional feedback system with $P(s) = k_{pll}/s$ (the plant is an integrator). The compensator appears after the plant, but the change does not affect the analysis. The objective is to have $x_{vco}(t)$ track the scaled signal $x_s(t)$, despite the constant disturbance signal originating from the difference in center frequencies between the modulator and the demodulator. The filter $C(s)$ is to be designed as a control system to achieve this result.

4.7.4   Compensator design

From Fig. 4.43, the Laplace transforms $X_{vco}(s)$, $\Phi(s)$, $X_s(s)$, $D(s)$ are related by

$$X_{vco}(s) = H_x(s)\,(X_s(s) + D(s)), \qquad \Phi(s) = H_\phi(s)\,(X_s(s) + D(s)), \qquad (4.103)$$

where

$$H_x(s) = \frac{k_{pll} C(s)}{s + k_{pll} C(s)}, \qquad H_\phi(s) = \frac{k_{pll}}{s + k_{pll} C(s)}. \qquad (4.104)$$

Two typical choices of compensator $C(s)$ are:

- $C(s) = \dfrac{k_f}{s + a_f}$ first-order filter

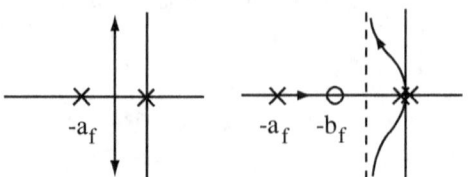

Figure 4.44: Root-locus for first-order filter (left) and second-order filter (right)

- $C(s) = \dfrac{k_f(s+b_f)}{s(s+a_f)}$ second-order filter

For the first-order filter,

$$H_x(s) \;=\; \frac{k_{pll}k_f}{s^2 + a_f s + k_{pll}k_f},$$

$$H_\phi(s) \;=\; \frac{k_{pll}(s+a_f)}{s^2 + a_f s + k_{pll}k_f} \qquad \text{(first-order).} \qquad (4.105)$$

The second-order filter includes an integrator, and yields the transfer functions

$$H_x(s) \;=\; \frac{k_{pll}k_f(s+b_f)}{s^3 + a_f s^2 + k_{pll}k_f s + k_{pll}k_f b_f},$$

$$H_\phi(s) \;=\; \frac{k_{pll}s(s+a_f)}{s^3 + a_f s^2 + k_{pll}k_f s + k_{pll}k_f b_f} \qquad \text{(second-order).} \qquad (4.106)$$

Fig. 4.44 on the left shows the root-locus for the first-order filter, for $k_{pll}k_f = 0 \to \infty$ and $a_f > 0$. On the right, the root-locus is shown for the second-order filter, assuming $a_f > b_f > 0$. With these restrictions on the parameters, the closed-loop system is always stable. Closed-loop poles may be placed at appropriate locations by proper choice of the parameters.

For both filters, the DC gain of the transfer function from x_s to x_{vco} is equal to 1. Therefore, $x_{vco}(t)$ (the frequency estimate) will match $x_s(t)$ (the true frequency) in steady-state, and $x_{vco}(t)$ will track $x_s(t)$, as long as $x_s(t)$ does not vary too fast compared to the time constants of the closed-loop system. The magnitude of the poles should be high enough to ensure tracking of the signal $x_s(t)$, but not so high to yield excessive sensitivity to noise.

Consider now the effect of a center frequency error δf_c. In the steady-state, the effect of δf_c on x_{vco} and ϕ is determined by $G_x(0)$ and $G_\phi(0)$. In the case of a first-order filter, a center frequency error δf_c results in a constant bias $\delta x_{vco,ss}$ and a constant phase error $\delta \phi_{ss}$ given by

$$\delta x_{vco,ss} = \frac{\delta f_c}{k_{vco}}, \qquad \delta \phi_{ss} = \frac{a_f}{k_f}\frac{\delta f_c}{k_{vco}} \qquad \text{(first-order).} \qquad (4.107)$$

The steady-state frequency error $f(t) - f_{vco}(t) = (1/2\pi)(d\phi/dt)$ is zero since ϕ is constant. Therefore, the signals y and y_{vco} have the same instantaneous frequencies. There is a phase difference $\delta\phi_{ss}$: although the two phases are locked, they are not equal. The phase difference produces a bias at the input of the VCO $(\delta x_{vco,ss})$, which ensures that the frequencies of the incoming and VCO signals are matched. Because the phase detector is only linear for some range of ϕ, the center frequency error should be small enough to operate in that region. For example, the steady-state phase $\delta\phi_{ss}$ will be smaller than π (or 180 degrees) if the center frequency error satisfies

$$\delta f_c < \frac{k_f}{a_f} k_{vco} \pi. \tag{4.108}$$

The center frequency of the PLL must be much closer to the center frequency of the modulator than the bound specifies for operation in the linear region to be possible.

With the second-order filter, the steady-state signals are

$$\delta x_{vco,ss} = \frac{\delta f_c}{k_{vco}}, \qquad \delta\phi_{ss} = 0 \qquad \text{(second-order)}. \tag{4.109}$$

Now, both the frequency error *and* the phase error are zero despite a center frequency error. The bias in the VCO input δx_{vco} remains the value required to match the incoming frequency, but it is now provided by the integrator of the compensator. The principle of operation is similar to the rejection of constant disturbances using integral compensation in conventional feedback systems.

4.7.5 Phase detectors

The implementation of the phase-locked loop requires a phase detector. An ideal phase detector was assumed for the linear analysis of the previous section. However, practical devices are less than ideal. A phase detector based on a multiplier is shown in Fig. 4.45.

The principle of operation is as follows. Given

$$y(t) = \sin(\theta(t)), \quad y_{vco}(t) = \sin(\theta_{vco}(t)), \tag{4.110}$$

the phase-advanced signal y_q is given by

$$y_q(t) = \cos(\theta_{vco}(t)), \tag{4.111}$$

and

$$2\,y(t)\,y_q(t) = \sin(\theta(t) - \theta_{vco}(t)) + \sin(\theta(t) + \theta_{vco}(t)). \tag{4.112}$$

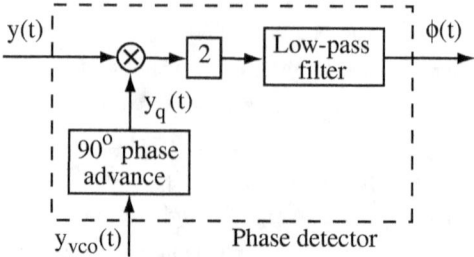

Figure 4.45: Phase detector

The second component of (4.112) is a signal whose frequency is approximately twice the center frequency of the incoming signal. The purpose of the low-pass filter is to eliminate that signal, so that

$$\phi(t) = LPF\left[2\ y(t)\ y_q(t)\right] \simeq \sin\left(\theta(t) - \theta_{vco}(t)\right). \tag{4.113}$$

Then, the desired result is obtained

$$\phi(t) \simeq \theta(t) - \theta_{vco}(t) \tag{4.114}$$

if $\theta(t) - \theta_{vco}(t)$ is small.

Regarding the 90° phase advance, a first option is to compute $\cos(\cdot)$ in the VCO of the PLL instead of $\sin(\cdot)$. A second option is to eliminate the phase advance block, recognizing that the steady-state phase $\theta_{vco}(t)$ will simply be offset by 90°, and that there will be no effect on the signal $x_{vco}(t)$. The steady-state signals $y(t)$ and $y_{vco}(t)$ will be in quadrature, rather than in phase.

The characteristics of the phase comparator are only ideal for small phase error $\theta(t) - \theta_{vco}(t)$. The overall behavior is nonlinear, as shown by (4.113) and on Fig. 4.46. With more sophisticated phase detectors, the characteristic is linear for a larger range of ϕ. However, the periodicity of 2π is unavoidable, since phases separated by 2π cannot be distinguished. Therefore, any practical phase comparator will be nonlinear, and at most linear up to an angle π. A precise analysis of the feedback system is much more complex than the linear approximation indicates.

From Fig. 4.46, one may note that $\phi = \pi$ is also an equilibrium state of the PLL with multiplier-based phase detector. However, it corresponds to a sign reversal of the feedback gain and to an unstable condition. Although phases separated by 2π cannot be distinguished, it is possible for the phase of two

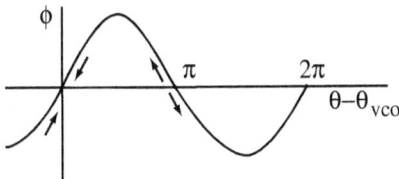

Figure 4.46: Nonlinear characteristic of the phase comparator

signals with nearly identical frequency to slowly increase from 0 and become 2π, 4π, etc. When there is an initial error in frequency $f_c - f_{c,vco}$ in a PLL, it may take several cycles for ϕ to converge to a stable equilibrium position. This condition is referred to as *cycle slipping*. In some cases, phase lock may never occur. The *lock-in range* is the range of frequencies for which phase lock occurs without cycle slipping. The *capture range*, or *pull-in range*, is the range of frequencies for which phase lock occurs, possibly with cycle slipping. One also defines the *hold range* as the range of frequencies for which the loop *remains* locked, if it is initially locked. This range may be determined experimentally by slowly varying the frequency of the incoming signal, starting from a locked condition.

The nonlinear behavior highlights two additional considerations in the choice of the closed-loop bandwidth of the linearized system: the bandwidth should be small enough to filter the high frequency component originating from phase detection, but the response should be fast enough to ensure locking of the PLL. In general, while the design of the loop filter is performed using linear time-invariant analysis methods, characteristics of the true nonlinear system must be considered as well.

4.8 Problems

Problem 4.1: Consider the control system of Fig. 4.47.
(a) Let $P(s) = k/(s + a)$ and $C(s) = k_P$. Find $Y(s)$, assuming that both $R(s)$ and $D(s)$ are nonzero. Deduce the transfer functions from r to y (for $d = 0$) and from d to y (for $r = 0$). Give conditions on k_P, k and a such that the transfer functions are BIBO stable. Assuming that such conditions are satisfied, give the values of the DC gains of the transfer functions. Indicate whether perfect tracking of constant reference inputs and perfect rejection of constant

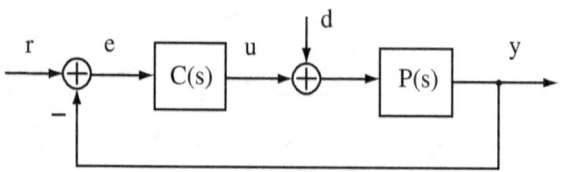

Figure 4.47: Standard feedback system with disturbance

disturbances is achieved by the control system.

(b) Consider the special case of a DC motor, with input u (the voltage in V) and output y (the speed in rad/s). Let $k = 1000$ (rad/s)^2/V, $a = 100$ rad/s, $k_P = 3$ V/(rad/s) (or V s), $r = 200$ rad/s and $d = 0$. Give the steady-state value of y and the steady-state value of the error e. Plot the steady-state error, expressed as a percentage of the reference input, as a function of k_P. Observe that the error goes to zero as $k_P \to \infty$. Also show that it is possible to achieve perfect tracking by multiplying the reference input by a constant number before it is applied to the summing junction. Explain the limitation of this approach to tracking.

(c) In the same conditions as part (b), except that $r = 0$ and $d = 3$ V (the disturbance is expressed as an equivalent input disturbance in volts), give the steady-state value of y and of the error e. Plot the error for a unit disturbance ($d = 1$) as a function of k_P. Show that it is possible to add to the reference input a signal that is proportional to the disturbance, so that the output is equal to the reference input. Explain the limitation of this approach to disturbance rejection.

(d) Repeat part (a) with $P(s) = k/(s+a)$ and $C(s) = k_P/s$.

(e) Repeat part (a) with $P(s) = k/(s(s+a))$ and $C(s) = k_P$.

Problem 4.2: Determine whether all the roots of the following polynomials are in the open left half-plane

(a) $D(s) = s^4 + 4s^3 + 3s^2 + 4s + 1$

(b) $D(s) = s^5 + 5s^4 + 8s^3 + 4s^2 - s - 1$

(c) $D(s) = s^4 + 2s^3 + 2s^2 + 2s + 1$

Problem 4.3: Give necessary and sufficient conditions on a and b for the following polynomial to have all roots in the open left half-plane: $D(s) = s^4 + s^3 + s^2 + as + b$.

Problem 4.4: Give conditions on k_P, k_I, and k_D such that a closed-loop system with $P(s) = \dfrac{k}{s(s+a)}$ and $C(s) = k_P + \dfrac{k_I}{s} + k_D s$ is BIBO stable. Assume that $k > 0$, but consider both cases $k_I \neq 0$ and $k_I = 0$.

Problem 4.5: (a) Sketch the root-locus of the transfer function

$$G(s) = \frac{s(s+1)}{(s+2)^2(s+3)} \tag{4.115}$$

applying only the main rules.

(b) Repeat part (a) for

$$G(s) = \frac{(s+3)}{s(s+9)^3}. \tag{4.116}$$

(c) Repeat part (a) for

$$G(s) = \frac{s+a}{(s+b)(s^2 - 2s + 2)}. \tag{4.117}$$

Also give condition(s) that $a > 0$ and $b > 0$ must satisfy for the closed-loop system to be stable for sufficiently high gain $k > 0$ (note that you do not need to apply the Routh-Hurwitz criterion, nor provide the range of k for which the system is closed-loop stable).

Problem 4.6: Sketch the root-locus for the transfer function

$$G(s) = \frac{(s+3)}{(s-1)(s^2 + 2s + 2)}. \tag{4.118}$$

Give the range of $k > 0$ for which the system is closed-loop stable and calculate the angle of departure of the locus from the pole at $s = -1 + j$.

Problem 4.7: Sketch the root-locus for the transfer function

$$G(s) = \frac{(s+1)^2}{s^3}. \tag{4.119}$$

Give the locations of the breakaway points on the real axis, the values of the angles of departure, and the range of $k > 0$ for which the system is closed-loop stable.

Problem 4.8: Sketch the root-locus for the transfer function

$$G(s) = \frac{1}{(s-1)(s+3)^2} = \frac{1}{s^3 + 5s^2 + 3s - 9}. \tag{4.120}$$

Give the range of gain $k > 0$ for which the closed-loop system is stable, the locations of the breakaway points on the real axis, and the locations of the $j\omega$-axis crossings.

Problem 4.9: Sketch the root-locus for the transfer function

$$G(s) = \frac{(s^2 + 2s + 17)}{(s+1)^3} = \frac{(s+1+4j)(s+1-4j)}{(s+1)^3}. \tag{4.121}$$

Include the angles of departure and arrival. Also give the range of gain $k > 0$ for which the closed-loop system is stable and use the result to improve your sketch of the locus, if possible.

Problem 4.10: Consider a standard control system as shown in Fig. 4.47. Let

$$P(s) = \frac{1}{s^2(s+1)}, \quad C(s) = k_P + \frac{k_I}{s} + k_D s. \tag{4.122}$$

(a) Assuming that $k_P \neq 0$ and that the closed-loop system is stable, what condition(s) must k_I and k_D satisfy so that the steady-state error for constant reference inputs is zero?

(b) Assuming that $k_P \neq 0$ and that the closed-loop system is stable, what condition(s) must k_I and k_D satisfy so that the steady-state error for constant input disturbances is zero?

(c) Assuming that $k_I \neq 0$, what condition(s) must k_P, k_I, and k_D satisfy so that the closed-loop system is stable?

Problem 4.11: (a) Sketch the root-locus for $k > 0$ and the problem of Fig. 4.48. There is one zero at $s = 0$ and two poles at $s = 1$.

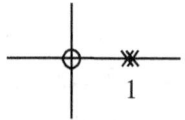

Figure 4.48: Pole/zero locations for problem 4.11

(b) Give the range of gain $k > 0$ for which the system is closed-loop stable, and give the locations of the $j\omega$-axis crossings.

(c) Give the locations of the breakaway points on the real axis.

Problem 4.12: Sketch the root-locus for the problem of Fig. 4.49, using only the main rules. There is a zero at $s = 0$, two poles at $s = -1$, and two poles at $s = -1 \pm j$.

Problem 4.13: Sketch the root-locus for the problem of Fig. 4.50. Do not calculate the range of gains for stability, the $j\omega$-axis crossings, or the breakaway points from the real axis. However, give the angles of departure from the complex

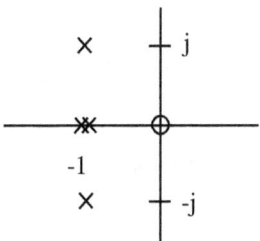

Figure 4.49: Pole/zero locations for problem 4.12

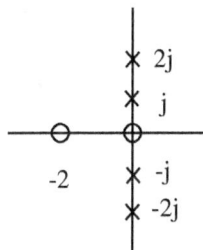

Figure 4.50: Pole/zero locations for problem 4.13

poles. There is a zero at $s = 0$ and a zero at $s = -2$. There are poles at $s = \pm j$ and $s = \pm 2j$. Note that $\tan^{-1}(0.5) = 27°$.

Problem 4.14: Determine whether the roots of the polynomial $D(s)$ are all in the open left half-plane, where

$$D(s) = s^5 + 3s^4 + 4s^3 + 4s^2 + 3s + 1. \tag{4.123}$$

Problem 4.15: (a) Consider the system of Fig. 4.47. Let

$$P(s) = \frac{1}{s(s+1)}, \quad C(s) = \frac{s+a}{s+1}. \tag{4.124}$$

Determine the range of values of the parameter a of $C(s)$ such that the closed-loop system is stable.

(b) For the system of part (a), give the steady-state error e_{ss} that is observed when $r(t) = 2$ and $d(t) = 0$. The result may be a function of the parameter a.

(c) For the system of part (a), give the steady-state error e_{ss} that is observed when $r(t) = 0$ and $d(t) = 2$. The result may be a function of the parameter a.

Problem 4.16: (a) Sketch the root-locus for

$$G(s) = \frac{s+2}{s^2(s^2 + 2s + 2)}. \tag{4.125}$$

Calculate the angles of departure for the complex poles, but do not calculate the breakaway points from the real axis, the range of gain for stability, or the crossing points on the $j\omega$-axis.

(b) Sketch the root-locus for

$$G(s) = \frac{1}{(s-1)(s+5)^2}. \tag{4.126}$$

Calculate the breakaway points from the real axis and the range of gain for stability (the crossing points on the $j\omega$-axis are not needed).

Problem 4.17: Consider the feedback system of Problem 4.4. Assuming that the stability conditions are satisfied, determine what the steady-state error $e_{ss} = \lim_{t\to\infty} e(t)$ is for $r(t) = 2$. Consider both cases $k_I \neq 0$ and $k_I = 0$.

Problem 4.18: (a) Sketch the root-locus for

$$G(s) = \frac{k}{s^4 + 6s^3 + 13s^2 + 12s + 4} \tag{4.127}$$

using only the main rules (the poles are shown below). Give the range of gain $k > 0$ for which the system is closed-loop stable. Poles are shown on Fig. 4.51.

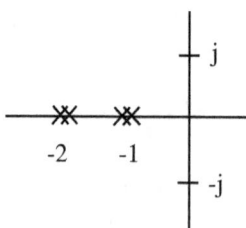

Figure 4.51: Pole/zero locations for problem 4.18 (a)

(b) Sketch the root-locus for the problem of Fig. 4.52, using only the main rules. Give the angles of departure from the complex poles.

Problem 4.19: (a) Find $Y(s)$ as a function of $R(s)$ and $D(s)$ for the system of Fig. 4.53.

(b) For the system of part (a), let $C_1(s) = 1/(s+a)$, $C_2(s) = k(s+1)/(s+2)$, $P(s) = 1/s$, and find conditions on k and a such that the closed-loop system is stable.

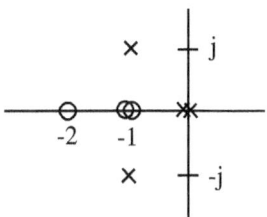

Figure 4.52: Pole/zero locations for problem 4.18 (b)

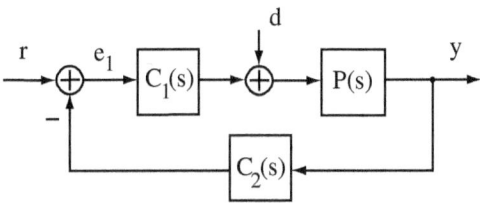

Figure 4.53: System for problem 4.19

(c) For the system of parts (a)-(b), let $e(t) = r(t) - y(t)$. Note that the signal is *not* equal to the signal denoted e_1 on the diagram. Assuming constant but arbitrary signals $r = r_0$ and $d = d_0$, obtain $E(s)$ and conditions on k and a such that $\lim_{t \to \infty} e(t) = 0$.

Problem 4.20: (a) Sketch the root-locus for

$$G(s) = \frac{1}{s((s+10)^2 + 4)}. \tag{4.128}$$

Compute the breakaway points from the real axis and the value of k for which crossing of the $j\omega$-axis occurs. Do not compute the angles of departure.

(b) Sketch the root-locus for

$$G(s) = \frac{1}{s((s+4)^2 + 16)}. \tag{4.129}$$

Compute the breakaway points from the real axis and the angles of departure from the complex poles. Do not compute the value of k for $j\omega$-axis crossing.

Chapter 5

Frequency-domain analysis of control systems

5.1 Bode plots

5.1.1 Motivation

While control systems may be designed on the basis of pole/zero knowledge, some applications are better handled using the frequency response of the system to be controlled. Indeed, the frequency response may be measured with good accuracy by injecting sinusoids of various frequencies at the input of the system. Given a system with transfer function $P(s)$, Bode plots are plots of the magnitude and of the angle of $P(j\omega)$ as a function of ω, *i.e.*, of the gain and phase shift of the steady-state sinusoidal response in the case of a stable system [5]. For example, Fig. 5.1 shows the magnitude Bode plot of a low-pass filter, as may be encountered in filtering applications.

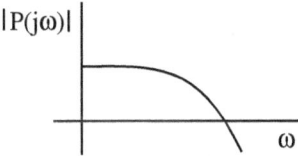

Figure 5.1: Magnitude Bode plot of a low-pass filter (log scales are used)

In the context of control systems, Bode plots are used even when the system is unstable, generalizing the concept of frequency response. The Bode plots are computed by replacing s by $j\omega$ in the transfer function $P(s)$. The frequency response does not exist for unstable systems, due to the lack of steady-state

sinusoidal response, but the Bode plots can nevertheless be determined experimentally by placing the system in a stabilizing feedback loop, as shown in
Fig. 5.2. In this manner, the effect of initial conditions will decay to zero and
the responses will remain bounded. A reference signal $r = r_0 \sin(\omega_0 t)$ is applied
and, assuming a stable closed-loop system, the signals u and y will converge to
the steady-state responses

$$
\begin{aligned}
u_{ss} &= u_0 \sin(\omega_0 t + \alpha_0), \\
y_{ss} &= y_0 \sin(\omega_0 t + \beta_0).
\end{aligned}
\tag{5.1}
$$

Then

$$
|P(j\omega_0)| = \frac{y_0}{u_0}, \quad \angle P(j\omega_0) = \beta_0 - \alpha_0.
\tag{5.2}
$$

In other words, the Bode plots can be measured and interpreted similarily for
stable and unstable systems. One just needs to remember that the steady-state
sinusoidal response only exists for unstable systems if the systems are placed in
a stabilizing feedback loop.

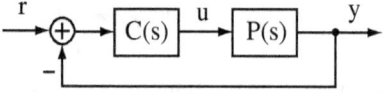

Figure 5.2: Closed-loop system

As for the root-locus, Bode plots can be computed easily and rapidly using
modern software. The procedures described in this chapter are not useful to
draw manually detailed plots, but rather to gain a valuable understanding of how
transfer function parameters are related to frequency response characteristics.

5.1.2   Approximations of the frequency response

Fig. 5.3 shows the Bode plots for the transfer function $G(s) = 1/(s + 1)$, with
the magnitude plot on the top and the phase plot on the bottom. Both plots
use a log scale on the x-axis, such that equal space is assigned to a range from
a given frequency to a frequency ten times greater. Such a frequency range is
referred to as a *decade*. The y-axis of the magnitude plot also uses a log scale,
but is labelled in dB (from the unit for sound level, the *decibel*), with

$$
[G(j\omega)]_{\mathrm{dB}} = 20 \log |G(j\omega)|.
\tag{5.3}
$$

On the y-axis, a decade spans a range of 20 dB. The phase plot shows $\angle G(j\omega)$ in a regular scale labelled in degrees.

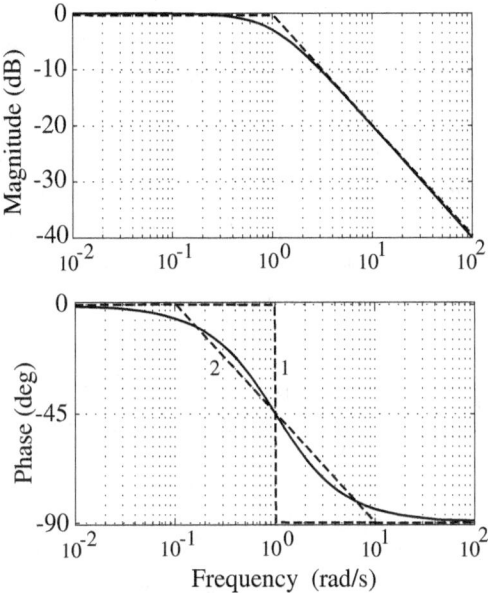

Figure 5.3: Bode plots and approximations

The plots on Fig. 5.3 show the magnitude and phase responses as solid lines and approximations as dashed lines. The magnitude approximation is very close. In the phase plot, two approximations are shown, with a coarse one labelled #1 and a finer one labelled #2. For the transfer function $G(s) = 1/(s+1)$, the approximations are based on the fact that $G(j\omega) \simeq 1$ for $\omega \ll 1$ and $G(j\omega) \simeq 1/(j\omega) = -j/\omega$ for $\omega \gg 1$, resulting in the magnitude and phase approximations given in the table below.

$\omega \ll 1$	$\omega \gg 1$
$[G(j\omega)]_{\mathrm{dB}} \simeq 0$	$[G(j\omega)]_{\mathrm{dB}} \simeq -20\log(\omega)$
$\angle G(j\omega) \simeq 0$	$\angle G(j\omega) \simeq -90°$

In the case of the magnitude plot, the approximation is composed on two lines. The first line is flat at 0 dB/decade and the second line slopes in the negative direction at -20 dB/decade. The two lines intersect at a point with frequency $\omega = 1$ rad/s and magnitude equal to 0 dB. The transition frequency is equal to the magnitude of the pole at $s = -1$. For the phase plot, approximation #1 is a discontinuous curve composed of a flat line at 0 deg and another flat line at $-90°$ with a sharp transition at $\omega = 1$ rad/s. This approximation is quite coarse, and a finer approximation is shown with the dashed line labelled #2. Both approximations can be used to produce Bode plots, with the second one more accurate but also more time-consuming to apply.

A similar approximation is obtained for $G(s) = 1/(s + p)$ with $p > 0$, except that the transition occurs at $\omega = |p|$ and the low frequency magnitude is $-20\log(|p|)$. If $p < 0$, the phase plot is similar but moving from $-180°$ to $-90°$. For a zero, $G(s) = s + z$, the plot for the magnitude is similar but adding 20 dB/decade at $\omega = |z|$. The phase plots are the same as for poles, but with signs reversed. Overall, the changes occuring at the transition frequency are summarized below.

	OLHP pole	ORHP pole	OLHP zero	ORHP zero
Magnitude	-20 dB/dec	-20 dB/dec	20 dB/dec	20 dB/dec
Phase	$-90°$	$90°$	$90°$	$-90°$

Note that the coarse approximation is recommended, because it can obtained much faster, while precise Bode plots are easily computed numerically nowadays.

5.1.3   Bode plots - Systems with no poles or zeros at the origin

Procedure

Step 1: prepare to draw two plots. For both plots, the x-axis is the log of the frequency ω. Although a log scale is used, the x-axis is typically labelled in ω directly (rad/s or Hz), rather than $\log(\omega)$. A scale for the x-axis might be 0.1, 1, 10, 100, $\cdots$, (rad/s). A factor of 10 on the x-axis is called a decade, and corresponds to an addition of 1 in $\log(\omega)$. A good choice of x-axis normally spans from

$$\omega_{\min} = \frac{1}{10}\min_{i,j}\left(|z_i|,|p_j|\right) \quad \text{to} \quad \omega_{\max} = 10\max_{i,j}\left(|z_i|,|p_j|\right), \tag{5.4}$$

where z_i are the zeros of the system and p_j are the poles (whose values have the units of rad/s). The first plot shows the magnitude of the frequency response as a function of $\log(\omega)$. A log scale is used again, with $20\log|P(j\omega)|$ shown on the y-axis. The units are dB's. A multiplication of the magnitude by 10 translates into an addition of 20 dB. The second plot is a phase plot, giving the angle (in degrees) of $P(j\omega)$ as a function of $\log(\omega)$. A log scale is not used for the y-axis of this plot.

Step 2: start the plots on the left at a sufficiently low frequency, such as ω_{min} in (5.4). For the magnitude, draw a horizontal line at $20\log(|P(0)|)$. For the phase, draw a horizontal line at $0°$ if $P(0) > 0$ and $180°$ if $P(0) < 0$.

Step 3: continue the plots from left to right. Every time a pole or zero is encountered, that is, every time $\omega = |p_i|$ or $\omega = |z_i|$:

(a) for the magnitude, change the slope by an additional -20 dB/decade every time a pole is encountered, and 20 dB/decade every time a zero is encountered.

(b) for the phase, add $-90°$ whenever a left half-plane pole or right half-plane zero is encountered, and $90°$ whenever a right half-plane pole or left half-plane zero is encountered.

Step 4: draw smooth curves that fit the approximations.

Comments

In simple terms, the procedure amounts to:

1. start the Bode plots from the left using the low-frequency approximation $P_{LF}(s) \simeq P(0)$ for $s = j\omega$ small.

2. move from left to right, applying -20 dB/dec to the magnitude plot when a pole is reached and $+20$ dB/dec when a zero is reached.

3. for the phase plot, adding $-90°$ when an OLHP pole or ORHP zero is reached, and $+90°$ when an ORHP pole or OLHP zero is reached.

Example 1: $P(s) = \frac{s+1}{s+10}$. $|P(0)| = 0.1$ (or -20 dB) and $\angle P(0) = 0°$. Start the Bode plot around $\omega = 0.1$ ($\frac{1}{10} \times 1$) with this approximation. Next, move from left to right and apply the changes when the poles at $s = -1$ and $s = -10$ are encountered. The result is shown in Fig. 5.4. The dashed curves are smooth estimates of the frequency response, based on the linear approximations. From these estimates, one may predict, for example, that the response of the system to a $\cos(10t)$ input is a signal $M\cos(10t + \phi)$, with M slightly smaller than 1 and a phase advance ϕ around $45°$.

Example 2: $P(s) = \frac{s+10}{s+1}$ (similar to example 1, but the pole and the zero are reversed). The plot is shown on Fig. 5.5. Note that the output lags behind the input when the pole is smaller than the zero in magnitude.

Example 3: $P(s) = \frac{-s+10}{s+1}$ (similar to example 2, but with a right half-plane zero). The plot is shown on Fig. 5.6. Note that the phase lag is larger than in the previous example, due to the zero being in the right half-plane.

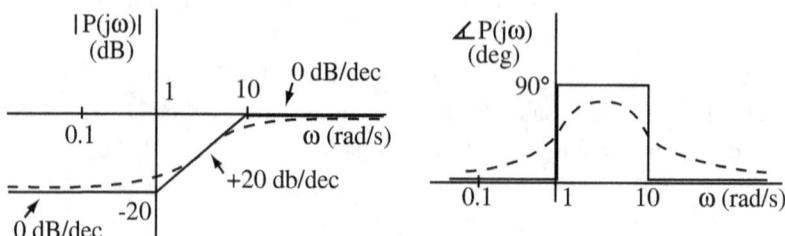

Figure 5.4: Bode plot for example 1

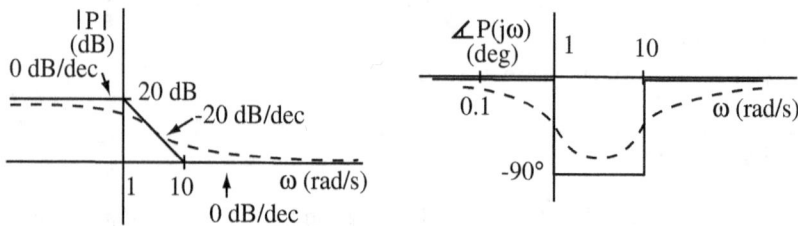

Figure 5.5: Bode plot for example 2

Such a zero is called a *non-minimum-phase* zero. One also says that a system is *minimum-phase* if all its zeros are in the left half-plane, and *non-minimum-phase* otherwise.

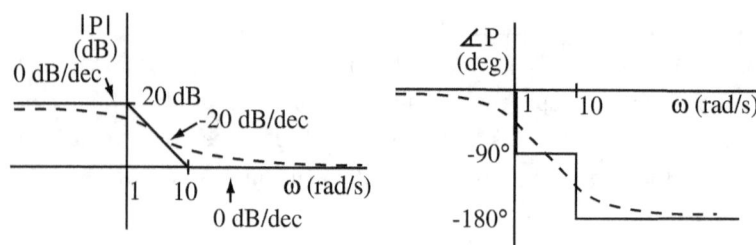

Figure 5.6: Bode plot for example 3

Example 4: $P(s) = \frac{s-10}{s+1}$. Here, $|P(0)| = 10$ (or 20 dB) and $\angle P(0) = 180°$. The plot is shown on Fig. 5.7.

Example 5: $P(s) = 10\frac{(s-1)}{(s-3)(s-10)}$. $|P(0)| = 1/3 \simeq -10$ dB and $\angle P(0) = 180°$. Note that the frequency 3 rad/s is approximately mid-way between 1 and

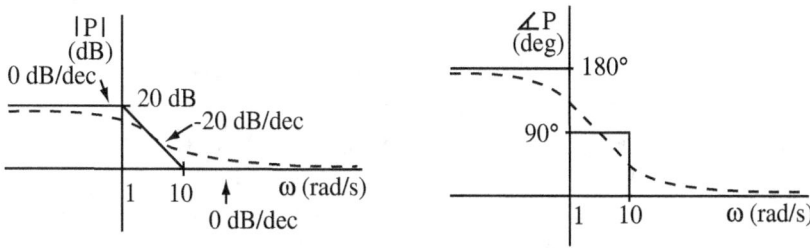

Figure 5.7: Bode plot for example 4

10 rad/s in a log scale ($\log(3) = 0.48$). The plot is shown on Fig. 5.8.

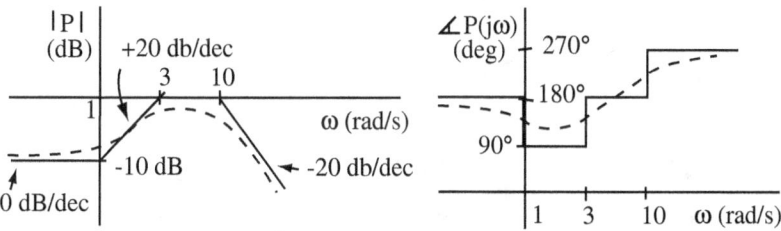

Figure 5.8: Bode plot for example 5

5.1.4 Bode plots - Systems with poles or zeros at the origin

Preliminary

The procedure remains the same, except that the low-frequency approximation in step 2 is more complicated. If the system has n zeros at $s = 0$, the low-frequency approximation is of the form

$$P_{LF}(s) = ks^n \quad \text{with } k = \left[s^{-n}P(s)\right]_{s=0}. \tag{5.5}$$

In other words, k is the DC gain of the transfer function with the zeros at the origin removed. If the system has n poles at $s = 0$, the same approximation applies with $n < 0$.

Procedure

In step 2, draw the low-frequency approximation as follows. Let $k = \left[s^{-n}P(s)\right]_{s=0}$ where n is the number of zeros of $P(s)$ at the origin (if there are poles at the origin instead, let $-n$ be the number of poles). For the magnitude, draw a line with a slope equal to $n \times 20$ dB/decade. Position the line so that $|P(j\omega_0)|_{dB} = 20 \log(|k|\omega_0^n)$, where ω_0 is some low frequency where the graph is started (for example, $\omega_0 = \omega_{min}$ using (5.4) computed from the poles and zeros other than those at the origin). For the phase, draw a horizontal line at $0° + n \, 90°$ if $k > 0$ and $180° + n \, 90°$ if $k < 0$.

Example 6: $P(s) = \frac{s+10}{s}$. Since $n = -1$, $k = 10$, the low-frequency approximation is $P_{LF}(s) = \frac{10}{s}$. Thus, we begin the plot using

$$|P_{LF}(j\omega)|_{dB} = \left|\frac{10}{j\omega}\right|_{dB} = 20 - 20\log(\omega). \tag{5.6}$$

At $\omega = 1$, $|P_{LF}(j\omega)|_{dB} = 20$ dB, which allows us to "pin down" the low-frequency approximation on the plot. The low-frequency phase is $\angle P(j\omega) = \angle(1/j) = -90°$. The plot is shown on Fig. 5.9.

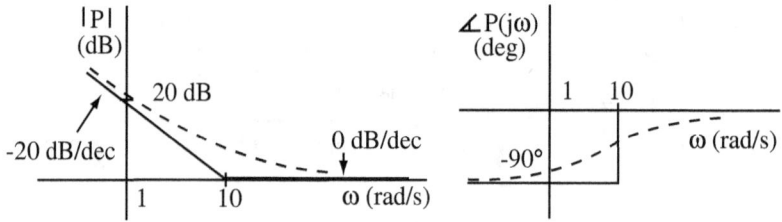

Figure 5.9: Bode plot for example 6

Example 7: $P(s) = \frac{s-10}{s^3}$. Since $n = -3$, $k = -10$, $P_{LF}(s) = \frac{-10}{s^3}$. We start at $\omega = 1$ with the approximation

$$|P_{LF}(j\omega)|_{dB} = \left|\frac{10}{\omega^3}\right|_{dB} = 20 - 60\log(\omega) \tag{5.7}$$

and $|P_{LF}|_{dB} = 20$ dB at $\omega = 1$. The low-frequency phase is given by

$$\angle P(j\omega) = 180° - 270° = -90°. \tag{5.8}$$

The plot is shown on Fig. 5.10.

5.1.5   Complex poles and zeros with low damping factor

Preliminary

Complex poles and zeros occur in complex pairs, with both poles or both zeros having the same magnitude. When a frequency ω is reached on the frequency

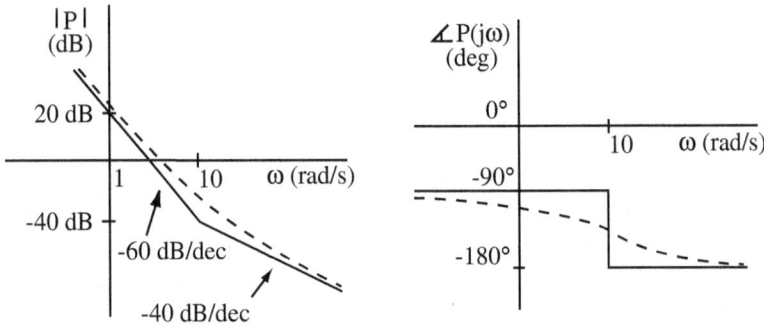

Figure 5.10: Bode plot for example 7

axis where $\omega = |p|$, the magnitude of the two poles or zeros, the effect is the same as if a pair of real poles or zeros had been reached (adding ± 40 dB/decade to the magnitude and $\pm 180°$ to the phase). For complex poles close to the $j\omega$-axis, however, the frequency response deviates significantly from the linear approximation.

Consider a system with a pair of stable complex poles $p = -a + jb$, $p^* = -a - jb$. One defines the natural frequency ω_n and the damping factor ζ through the formulas

$$\omega_n = |p| = \sqrt{a^2 + b^2},$$
$$\zeta = -\operatorname{Re}(p)/|p| = a/\sqrt{a^2 + b^2}. \tag{5.9}$$

These relationships are illustrated on Fig. 5.11. Note that $\zeta = \cos(\alpha)$, where α is the angle between the direction of the pole and the real axis.

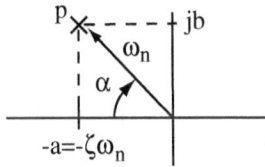

Figure 5.11: Definition of variables for complex poles

Consider the contribution of a complex pole pair to the transfer function

$$\frac{pp^*}{s^2 - (p + p^*)s + pp^*} = \frac{\omega_n^2}{s^2 + 2\zeta\omega_n s + \omega_n^2} \tag{5.10}$$

with the numerator set so that the DC gain is equal to 1. The Bode plot approximation is such that the gain is 1 (0 dB) up to ω_n, then decreases at the rate of -40 dB per decade. However, at ω_n, the exact gain is

$$|P(j\omega_n)| = \left|\frac{\omega_n^2}{2\zeta j\omega_n^2}\right| = \frac{1}{2\zeta}. \tag{5.11}$$

Therefore, the actual magnitude is 10 instead of 1 for $\zeta = 0.05$, or 20 dB. Fig. 5.12 shows the shape of the responses for different values of the damping factor ζ. For small damping factors, the magnitude response peaks significantly above the linear approximation. The transition of the phase response around ω_n is also sharper. For right half-plane poles, the phase is reversed and ζ is replaced by $|\zeta|$ for the plots.

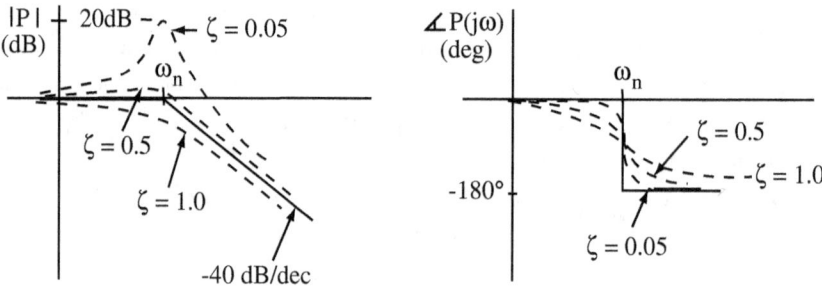

Figure 5.12: Bode plots of systems with two complex poles in the left half-plane

The true peak of the magnitude response is not exactly at ω_n, but at a frequency

$$\omega_p = \sqrt{1 - 2\zeta^2}\omega_n = \sqrt{b^2 - a^2}, \tag{5.12}$$

which is slightly smaller than ω_n. There is no peaking of the magnitude unless $\zeta < 0.707$, i.e., unless the imaginary part of the pole is greater than the real part. The actual magnitude of the peak is

$$|P(j\omega_p)| = \frac{1}{2\zeta} \frac{1}{\sqrt{1 - \zeta^2}}. \tag{5.13}$$

However, for small damping factor, the frequency of peaking is close to ω_n, and the magnitude is close to $1/2\zeta$, as given by (5.11).

For complex zeros close to the $j\omega$-axis in the open left half-plane, a similar correction needs to be applied and is shown in Fig. 5.13. For right half-plane

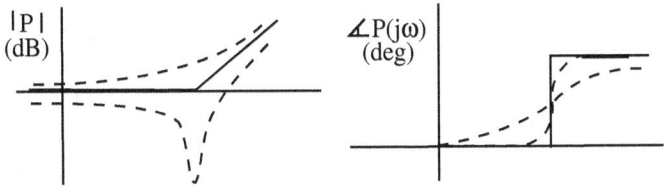

Figure 5.13: Bode plots for complex zeros in the left half-plane

zeros, the phase is reversed. If the zeros are on the $j\omega$-axis, the response is zero at the corresponding frequency, or at minus infinity in the log scale.

Procedure

When the frequency reaches a value equal to the magnitude of a pair of complex poles or zeros, apply the rules as if there were two real poles or two real zeros with the same magnitude. If the pair of complex poles or zeros is close to the $j\omega$-axis, peaking in the response may be accounted for as follows. Given a pair of complex poles $p = -a \pm jb$, let $\omega_n = \sqrt{a^2 + b^2}$ (called the *natural frequency*) and $\zeta = a/\omega_n$ (called the *damping factor*). If $|\zeta| < 0.5$, an increase of gain equal to $20\log(|1/2\zeta|)$ should be added to the magnitude plot at ω_n. For a pair of complex zeros, a similar reduction in the gain should be applied. The effect of a small damping factor on the phase plot is a rapid variation of phase around ω_n.

Example 8: Consider the transfer function

$$P(s) = \frac{s^2}{(s+1)\,(s^2 + 0.1s + 100)}, \tag{5.14}$$

which has poles at $p_1 = -1$, $p_{2,3} = -0.05 \pm \sqrt{(0.05)^2 - 100} = -0.05 \pm 9.999875$. The complex poles are such that $\omega_n = 10$, and $\zeta = 0.005$. Thefore, the peak of the gain at the natural frequency is $1/2\zeta = 100 \equiv 40$ dB. To draw the Bode plot, note that the low-frequency approximation is

$$P_{LF}(s) = \frac{s^2}{100}, \quad |P_{LF}(j\omega)|_{dB} = -40 + 40\log|\omega|, \quad \angle P_{LF}(j\omega) = 180°, \tag{5.15}$$

and, for $\omega = 0.1$,

$$|P_{LF}(0.1j)|_{dB} = -80 \text{ dB}. \tag{5.16}$$

The resulting Bode plot is shown in Fig. 5.14. The peak of the magnitude plot and the sharp phase transition at the complex poles were accounted for in the drawing of the smooth approximation.

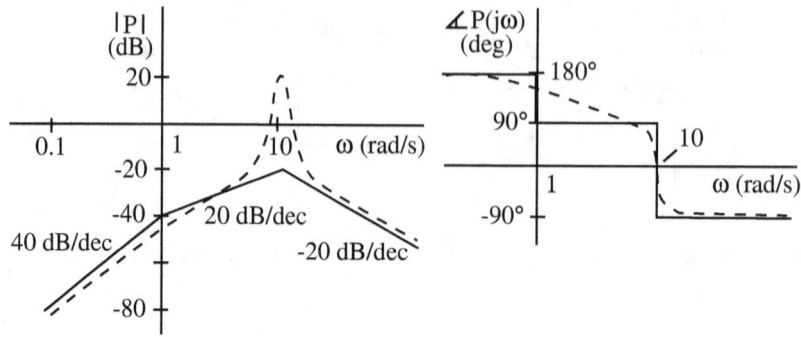

Figure 5.14: Bode plots of example with low damping complex poles

For $\omega = 10$ rad/s, the estimate of the Bode plot is $|P| = 20$ dB and $\angle P = 0°$, i.e., $P(j10) = 10$. The true value is

$$P(j10) = \frac{-100}{(1 + j10)(j)} = 9.9010 + 0.9901j, \tag{5.17}$$

which is close to the estimate.

5.1.6   Some special transfer functions

We discuss a few additional transfer functions that are commonly encountered, and their associated Bode plots.

Time delay: the transfer function of a time delay T_d (in seconds) is

$$P(s) = e^{-sT_d}. \tag{5.18}$$

Note that this is *not* a rational function of s, so that the usual rules of Bode plots cannot be applied. However, the plots themselves can be drawn. The frequency response of a time delay is $P(j\omega) = e^{-j\omega T_d}$ and

$$
\begin{aligned}
|P(j\omega)| &= 1, \\
\angle P(j\omega) &= -\omega T_d.
\end{aligned}
\tag{5.19}
$$

The Bode plots are shown in Fig. 5.15. A plot of the phase with a *linear* scale for the x-axis was inserted, to illustrate the property of the delay called *linear phase*. This linear phase property is associated with the fact that a time delay does not alter the shape of the signal (the signal is not distorted, as it generally would be with a rational transfer function).

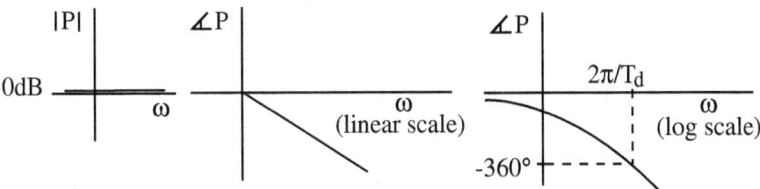

Figure 5.15: Bode plots of pure delay system

Notch filter: a notch filter eliminates the component of a signal at a given frequency ω_0. The transfer function of a second-order notch filter is given by

$$P(s) = \frac{s^2 + \omega_0^2}{s^2 + 2\zeta\omega_0 s + \omega_0^2}. \tag{5.20}$$

For example, one may let $\zeta = 1$. The Bode plots of the notch filter are shown in Fig. 5.16, together with the individual contributions of the complex poles and zeros. Note that the zeros are exactly on the $j\omega$-axis, so that the gain for a sinusoidal signal of frequency ω_0 is exactly zero ($P(j\omega_0) = 0$). Outside a narrow band around ω_0, $P(j\omega) \simeq 1$.

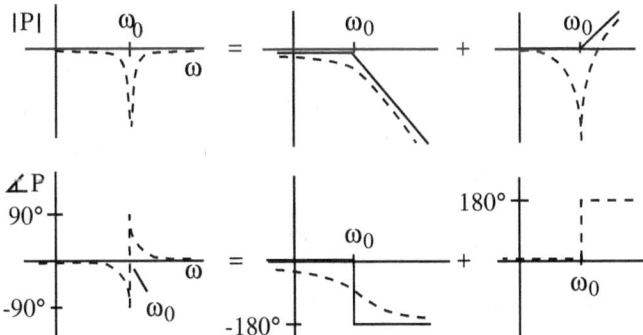

Figure 5.16: Bode plot of a notch filter, showing also the individual contributions of the denominator and numerator of the transfer function

Wash-out filter: a wash-out filter eliminates the DC component of a signal. An example of a wash-out filter is

$$P(s) = \frac{s}{s + a}. \tag{5.21}$$

It is a special case of a high-pass filter. The Bode plot of the filter is shown in Fig. 5.17. Wash-out filters are used to:

- eliminate biases and offsets in signals.

- keep the response of a system centered around a neutral position (an example is a flight simulator which emulates the motion of an aircraft, yet must remain at the same location).

- approximate the derivative of a signal within a finite frequency range (as in a PID controller).

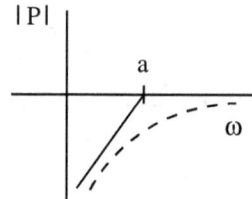

Figure 5.17: Magnitude Bode plot of a wash-out filter

Low-pass filter: the wash-out filter is a *high-pass* filter, because it filters out the low frequencies and lets high frequencies pass through. Conversely, a low-pass filter removes high-frequency components while transmitting the low-frequency components. For example, consider the signal of Fig. 5.18, meant to represent a noisy sinusoidal signal and given by

$$u(t) = \sin(2\pi t) + 0.3 \ \sin(2\pi \ 20 \ t). \tag{5.22}$$

A low-pass filter can be used to isolate the signal at 1 Hz, while reducing the noisy component at 20 Hz.

An example of low-pass filter is

$$F(s) = \frac{a^n}{(s+a)^n}, \tag{5.23}$$

where a determines the bandwidth of the filter, and n is the order of the filter. Fig. 5.19 shows the Bode plots of $F(s)$ for $a = 30$, and for $n = 1$ and $n = 2$. The magnitude plot shows the attenuation of high-frequency signals, which is enhanced for the higher value of n. Lower values of a also reduce high frequency

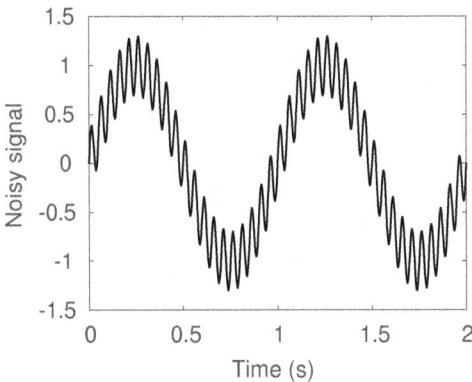

Figure 5.18: Noisy signal to demonstrate low-pass filtering.

components, but affect the main component as well if the value is too small. In the example, the main component is at $\omega = 6.28$ rad/s, while the noise is at $\omega = 125.7$ rad/s.

Fig. 5.20 shows the noisy signal filtered by $F(s)$ with $n = 1$ and $n = 2$, together with the original signal $\sin(2\pi t)$. For $n = 1$, the noise is considerably reduced, while it is virtually eliminated for $n = 2$. Notable features are that the main component is delayed compared to the original signal, and slightly reduced in magnitude as well. This effect is stronger for $n = 2$, and can be predicted from the magnitude and phase plots of Fig. 5.19. In general, the higher the reduction of the high-frequency noise, the greater the delay of the low frequency signal.

A useful observation is that the effect of the filter at low frequencies can be approximated by a pure time delay. Indeed, the transfer function of the filter (5.23) can be represented by the first terms of its Taylor series expansion around $s = 0$, with

$$F(s) \simeq 1 - \frac{n}{a}s + \ldots \tag{5.24}$$

Similarly, a time delay can be approximated by

$$F(s) = e^{-sT_d} \simeq 1 - T_d s. \tag{5.25}$$

Combining the expressions, one obtains the equivalent time delay of the low-pass filter

$$T_d = \frac{n}{a}. \tag{5.26}$$

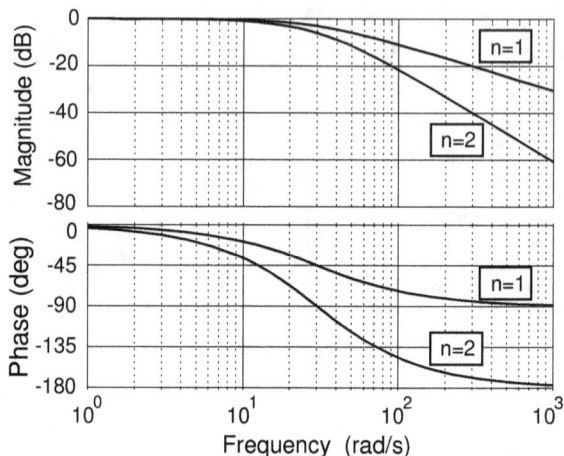

Figure 5.19: Bode plots of $a^n/(s+a)^n$ for $a = 30$, and for $n = 1$ and $n - 2$

Lower bandwidth or higher order filters imply higher time delays. In the example of Fig. 5.20, the approximate time delay $T_d = 33.3$ ms for $n = 1$, and $T_d = 66.7$ ms for $n = 2$.

Low-pass filters are often used to reduce the noise re-injected in the feedback loop from measurements of the output. However, the delay is detrimental to the closed-loop stability of the system, so that the choice of filter represents a trade-off between removal of the noise and closed-loop dynamics.

The transfer function (5.23) is only one example of a low-pass filter. Many low-pass filters exist, such as Butterworth filters, Bessel filters, Chebyshev filters, and elliptic filters. In general, the formula for the equivalent delay of a filter is

$$T_d = \lim_{s \to 0} \frac{F(0) - F(s)}{sF(0)}. \tag{5.27}$$

In the case of a stable transfer function

$$F(s) = \frac{b_{n-1}s^{n-1} + \cdots + b_1 s + b_0}{s^n + a_{n-1}s^{n-1} + \cdots + a_1 s + a_0}, \tag{5.28}$$

the formula gives

$$T_d = \frac{a_1}{a_0} - \frac{b_1}{b_0}. \tag{5.29}$$

The result requires that $a_0 \neq 0$ (needed for stability) and $b_0 \neq 0$ (needed for non-vanishing response at low frequencies).

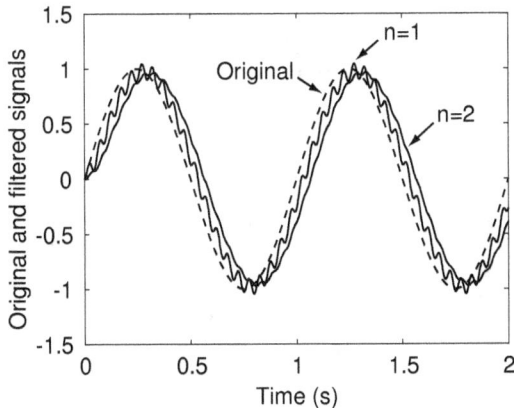

Figure 5.20: Original (dashed) and filtered (solid) signals

5.2   Nyquist criterion of stability

5.2.1   Nyquist diagram

The Nyquist criterion [23] is an important test to determine the stability of a closed-loop system, based on properties of the frequency response of the open-loop system. It may be applied using experimentally measured frequency response data, without obtaining a pole/zero model. The criterion is based on the Nyquist diagram, which is explained first.

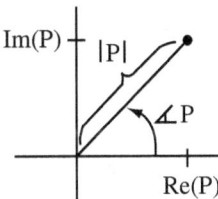

Figure 5.21: Polar representation of the frequency response

The Nyquist diagram is a plot of $\operatorname{Im} P(j\omega)$ vs. $\operatorname{Re} P(j\omega)$ for $\omega = -\infty \to \infty$. As shown in Fig. 5.21, it may also be viewed as a polar plot of the frequency response. Consider for example $P(s) = 1/(s+1)$. We have that

$$\text{Re}\,P(j\omega) = \frac{1}{1+\omega^2}, \quad \text{Im}\,P(j\omega) = \frac{-\omega}{1+\omega^2}. \tag{5.30}$$

Because

$$\left(\text{Re}\,P(j\omega) - \frac{1}{2}\right)^2 + (\text{Im}\,P(j\omega))^2 = \frac{\left(\frac{1}{2} - \frac{1}{2}\omega^2\right)^2 + \omega^2}{(1+\omega^2)^2} = \frac{1}{4}, \tag{5.31}$$

the Nyquist curve is a circle with radius $1/2$ and centered at $(1/2, 0)$. The diagram for positive frequencies is shown in Fig. 5.22. The arrow shows the direction for $\omega = 0 \rightarrow \infty$. If we consider instead $P(s) = 1/(s+1)^3$, the phase is multiplied by 3, and the resulting Nyquist diagram for positive frequencies is shown in Fig. 5.23.

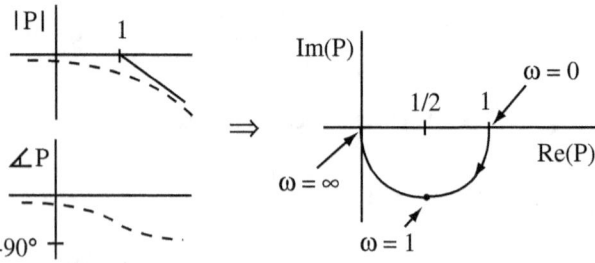

Figure 5.22: Bode plots (left) and Nyquist diagram (right) for $1/(s+1)$

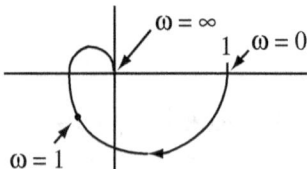

Figure 5.23: Nyquist diagram of a third-order system

The plot for $\omega > 0$ is complemented by the portion for $\omega < 0$. However, the fact that $P(-j\omega) = P^*(j\omega)$ (a consequence of the assumption that the impulse response is real) implies that the diagram for $\omega < 0$ is the reflection of the diagram for $\omega > 0$ with respect to the real axis. Fig. 5.24 shows the complete diagrams for the two examples.

Other properties worth noting are that:

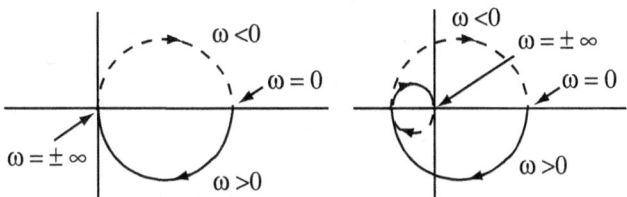

Figure 5.24: Complete Nyquist diagram for $1/(s+1)$ (left) and $1/(s+1)^3$ (right)

- If $P(s)$ has no pole at $s = 0$, $P(0)$ is real and finite ($P(0)$ is the DC gain of $P(s)$).

- If the number of poles is equal to the number of zeros, $P(\infty) = P(-\infty)$.

- If the number of poles is greater than the number of zeros, $P(\infty) = P(-\infty) = 0$.

- Assuming that $P(s)$ is of the form

$$P(s) = k\frac{s^m + \cdots}{s^n + \cdots},$$ (5.32)

with $n \geqslant m$ (proper transfer function), the high-frequency behavior may be approximated by

$$P_{HF}(j\omega) \simeq \frac{k}{\omega^{n-m}}e^{-j(\pi/2)(n-m)},$$ (5.33)

so that the Nyquist diagram approaches the origin for high frequencies with an angle

$$\begin{aligned} \angle P_{HF}(j\omega) &= -\frac{\pi}{2}(n - m) & \text{if } k > 0 \\ &= \pi - \frac{\pi}{2}(n - m) & \text{if } k < 0. \end{aligned}$$ (5.34)

5.2.2   Nyquist criterion

The Nyquist criterion considers both the *open-loop transfer function* $G(s) = N(s)/D(s)$ and the *closed-loop transfer function*

$$\frac{G(s)}{1 + G(s)} = \frac{N(s)}{N(s) + D(s)}.$$ (5.35)

The open-loop poles are the roots of $D(s) = 0$, and the closed-loop poles are the roots of $N(s) + D(s) = 0$. For simplicity, we will assume that there are no open-loop poles or closed-loop poles on the $j\omega$-axis. We will also assume that the number of poles is greater than or equal to the number of zeros.

Nyquist criterion

1. Plot $G(j\omega)$ for $\omega = -\infty \to \infty$.

2. Let N be the number of clockwise encirclements of $(-1, 0)$, that is, the number of times that the closed curve drawn by $G(j\omega)$ as $\omega = -\infty \to \infty$ encircles the $(-1, 0)$ point in the clockwise direction.

3. Then: $Z = N + P$, where:

$$Z = \text{number of closed-loop poles in the right half-plane}$$
$$P = \text{number of open-loop poles in the right half-plane}$$

Example 1: let $G(s) = 1/(s+1)$, $D(s) + N(s) = s + 2$. The Nyquist plot is shown in Fig. 5.25. There are no encirclements, so that $N = 0$. Since the system is open-loop stable, $P = 0$. Therefore, $Z = 0$, and the test confirms that the system is closed-loop stable. If we let $G(s) = k/(s+1)$, $D(s) + N(s) = s+1+k$, the diagram is simply expanded by a factor k. The number of encirclements does not change, and since the open-loop system remains stable, the test verifies that the closed-loop system is stable for all k.

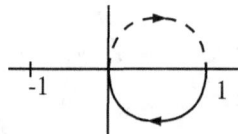

Figure 5.25: Nyquist diagram for $1/s + 1$

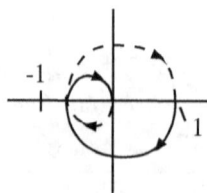

Figure 5.26: Nyquist diagram for $1/(s+1)^3$

Example 2: let $G(s) = 1/(s+1)^3$, $D(s) + N(s) = (s+1)^3 + 1$. Now, the Nyquist diagram is shown in Fig. 5.26. Since $N = 0$, $P = 0$, we have $Z = 0$, and the system is closed-loop stable. However, if we let $G(s) = k/(s+1)^3$, $D(s) + N(s) = (s+1)^3 + k$, there is a value of k such that the number of encirclements will change. The effect of an increasing parameter k on the Nyquist diagram and on the root-locus is shown in Fig. 5.27. When the value of k is sufficiently large that the value of a shown on the figure is greater than 1, the number of clockwise encirclements becomes equal to 2, and the number of unstable closed-loop poles also becomes $Z = 2$ (as indicated independently by the root-locus). We had found earlier using the Routh-Hurwitz criterion (see (4.46)) that the value of k which separated the stable and unstable conditions was $k = 8$. Considering the Nyquist criterion, we note that

$$G(j\omega) = \frac{k}{(1+j\omega)^3} = \frac{k}{(1+\omega^2)^3}\left((1-3\omega^2) + j\left(-3\omega + \omega^3\right)\right). \tag{5.36}$$

Therefore,

$$\mathrm{Im}\, G(j\omega) = 0 \quad \text{for} \quad -3\omega + \omega^3 = 0, \text{ or } \omega^2 = 3. \tag{5.37}$$

The crossing of the real axis by the Nyquist curve therefore occurs for $\omega^2 = 3$ and

$$a = -\mathrm{Re}\, G(j\omega) = -\frac{k(1-9)}{4^3} = \frac{k}{8}. \tag{5.38}$$

Again, one finds that the system is becomes unstable when $k > 8$ (the case for $k < 0$ can also be considered using the root-locus and the Nyquist criterion, but is left to the reader).

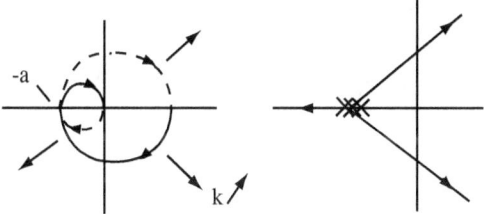

Figure 5.27: Nyquist diagram of $k/(s+1)^3$ for k increasing (left), and corresponding root-locus (right)

Example 3: $G(s) = k/(s-1)$ $(k > 0)$. This is an unstable open-loop system. Using the root-locus technique, as in Fig. 5.28, we can predict that the system is stable for sufficiently large $k > 0$. Indeed, since $D(s) + N(s) = s - 1 + k$, the system is known to be stable for $k > 1$. Drawing the Bode plots, we can sketch the Nyquist diagram as in Fig. 5.29. Because there is one unstable open-loop pole, $P = 1$. Then, for $k < 1$, $N = 0$, $Z = 1$ (1 unstable closed-loop pole). For $k > 1$, $N = -1$ (1 counterclockwise encirclement), $Z = 0$ (no unstable closed-loop pole). In summary, the Nyquist criterion correctly predicts that there is one unstable pole for $k < 1$, and that the closed-loop system is stable for $k > 1$.

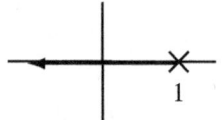

Figure 5.28: Root-locus of $k/(s - 1)$

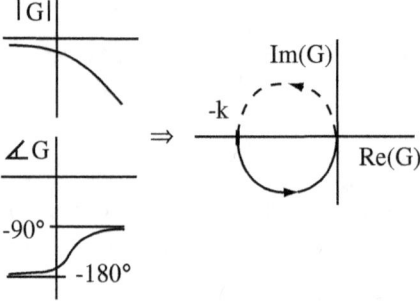

Figure 5.29: Bode plots (left) and Nyquist diagram (right) for $k/(s - 1)$

Example 4: consider the system

$$G(s) = \frac{k(s + 2)}{(s - 1)(s^2 + 2s + 2)} \tag{5.39}$$

The system has one unstable open-loop pole ($P = 1$) and its Nyquist curve shown in Fig. 5.30. The following results apply

$$
\begin{array}{llll}
k < 1 & N = 0 & Z = N + P = 1 \Rightarrow \text{unstable} \\
1 < k < 2 & N = -1 & Z = N + P = 0 \Rightarrow \text{stable} \\
k > 2 & N = 1 & Z = N + P = 2 \Rightarrow \text{unstable}
\end{array}
$$

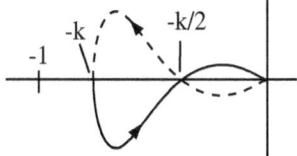

Figure 5.30: Nyquist diagram of a conditionally stable system

Such a system is called *conditionally stable:* the gain k is required to belong to a finite range for stability. The root-locus is shown in Fig. 5.31. While one pole becomes stable for $k > 1$, the other two poles become unstable when $k > 2$.

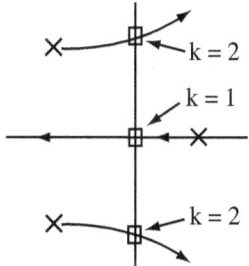

Figure 5.31: Example of root-locus for a conditionally stable system

5.2.3   Counting the number of encirclements

In some cases, counting the number of encirclements can be confusing. However, a simple procedure eliminates any ambiguity. As shown in Fig. 5.32 on the left, one may draw a straight line from $(-1, 0)$, and count the number of times that the Nyquist curve intersects the line (letting the sign be positive if the crossing is in the clockwise direction, and negative otherwise). Then, the sum is the number of clockwise encirclements. The number is zero on Fig. 5.32. As shown on the right of the figure, the number obtained is independent of the line that is used for the counting.

5.2.4   Implications of the Nyquist criterion

1. For the closed-loop system to be stable $(Z = 0)$, one must have:

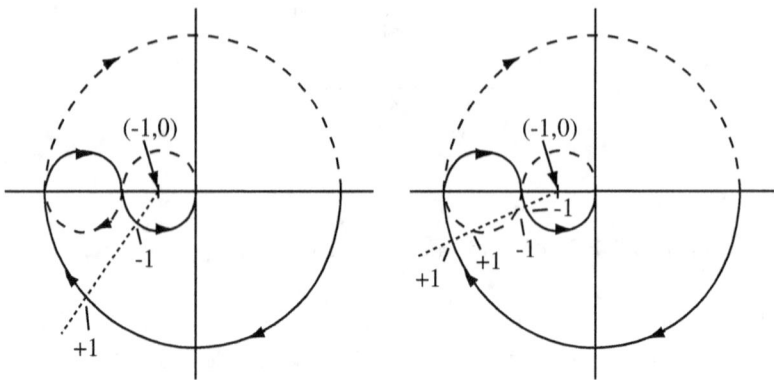

Figure 5.32: Counting the number of clockwise encirclements

(a) no encirclements if the open-loop system is stable.

(b) otherwise, as many *counterclockwise* encirclements as there are unstable open-loop poles ($N = -P$).

2. Note that, if the open-loop system is stable, *sufficient* (but not necessary) conditions for the stability of the closed-loop system are:

   - either $|G(j\omega)| < 1$ for all ω,

   - or $|\angle G(j\omega)| < 180°$ for all $\omega > 0$,

   - or, for some ω_1, $|\angle G(j\omega)| < 180°$ for $\omega \leqslant \omega_1$ and $|G(j\omega)| < 1$ for $\omega \geqslant \omega_1$.

3. One may wonder what happens when the Nyquist curve goes through $(-1, 0)$ point. Then, it is impossible to count the number of encirclements. However, $G(j\omega_0) = -1$ for some ω_0, so that $1 + G(j\omega_0) = 0$. Therefore, the system has at least one closed-loop pole at $j\omega_0$. This case was excluded in the assumptions, but one can nevertheless conclude that the closed-loop system is unstable, since some closed-loop poles are on the $j\omega$-axis.

4. The case where there are open-loop poles on the $j\omega$-axis was also excluded, but may be handled using a modified procedure to be described later. This case is important in feedback systems, because many control systems have poles at the origin.

5. The main value of the Nyquist criterion is to quantify how far a system is from being unstable. This topic will also be discussed later.

5.2.5   Explanation of the Nyquist criterion

Let C be the set of complex numbers located on a circle of radius 1 and associated with an orientation in the counterclockwise (CCW) direction, as shown on the left of Fig. 5.33. C is a closed curve and is also called a *contour*. Consider the transfer function

$$H(s) = s + z_1, \tag{5.40}$$

where z_1 is a real number. Note that $s = -z_1$ is the zero of $H(s)$. Let $H(C)$ be the set of complex numbers corresponding to $H(s)$ for $s \in C$ with an orientation corresponding to C. The right side of Fig. 5.33 shows $H(C)$, which is simply the original curve C shifted by z_1. A trivial fact is that $H(C)$ encircles the origin in the CCW direction if and only if the zero at $s = -z_1$ belongs to the interior of C. In the case shown on the figure, there are no encirclements.

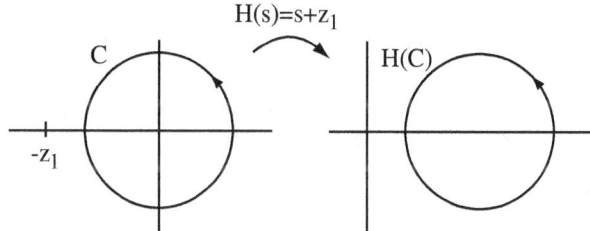

Figure 5.33: Contour and transformed contour

Fig. 5.34 shows an interpretation of the result where the complex number $H(s)$ is a vector with angle α with respect to the real axis. For $z_1 = 2$, as shown on the figure, and for $\angle s$ increasing from $0°$ to $360°$, α grows from 0 to $30°$, then goes down to $-30°$, then rises back to zero. $H(C)$ does not encircle the origin because the angle returns to $0°$ rather than reach $360°$. For $z_1 = 0$, α grows continuously from $0°$ to $360°$, and the origin is encircled.

With this interpretation, it becomes clear that the result on the number of encirclements remains true for arbitrary z_1 and for any closed curve C that does not intersect itself. Further, for a general transfer function

$$H(s) = b_m \frac{(s - z_1) \cdots (s - z_m)}{(s - p_1) \cdots (s - p_n)}, \tag{5.41}$$

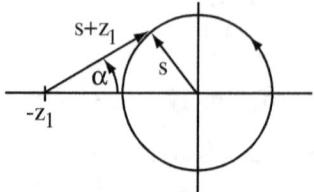

Figure 5.34: Angle of $H(s)$ as a function of the angle of s

one has that

$$\angle H(s) = \angle b_m + \sum_{i=1}^{m} \angle(s - z_i) - \sum_{i=1}^{n} \angle(s - p_i). \qquad (5.42)$$

It follows that

The number of CCW encirclements of the origin by $H(C)$ is equal to

the number of zeros of $H(s)$ inside C

$-$ the number of poles of $H(s)$ inside C $\qquad (5.43)$

The result is a version of the so-called *argument principle* or *Cauchy's principle of the argument* from complex analysis.

To obtain the Nyquist criterion, one lets

$$H(s) = 1 + G(s) = 1 + \frac{N(s)}{D(s)} = \frac{D(s) + N(s)}{D(s)}, \qquad (5.44)$$

where $G(s)$ is the open-loop transfer function. Therefore, the poles of $H(s)$ are the open-loop poles and the zeros of $H(s)$ are the closed-loop poles. Next, one defines the contour C as shown on Fig. 5.35. The contour is composed of the imaginary axis completed with a half circle of infinite radius. The curve is called the *Nyquist contour*. Note that the area of the complex plane delimited by the Nyquist contour is the right half-plane.

The orientation of the contour was changed to be clockwise (CW) so that the frequency varies in the positive direction along the imaginary axis. Instead of mapping $H(C) = 1 + G(C)$, one plots $G(C)$. Encirclement of the origin is replaced by encirclement of (-1, 0). If the loop transfer function is strictly proper, the half circle of infinite radius is mapped to Re = 0, Im = 0, and the Nyquist plot reduces to a plot of $G(j\omega)$ for ω ranging from $-\infty$ to ∞. With

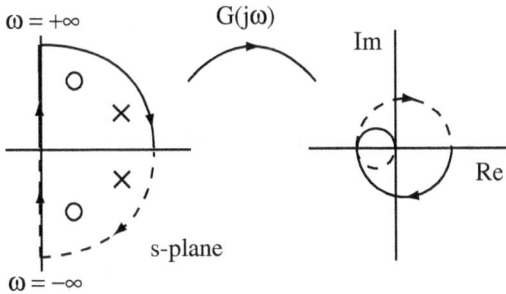

Figure 5.35: Principle of the Nyquist criterion

these changes, (5.43) becomes

The number of CW encirclements of (-1,0) by $G(j\omega)$ is equal to

the number of unstable closed-loop poles

$-$ the number of unstable open-loop poles,   (5.45)

which is the Nyquist criterion.

Whether the contour encompasses the right half-plane or the left half-plane is determined by the orientation of the contour. One could also have deduced that

The number of CCW encirclements of (-1,0) by $G(j\omega)$ is equal to

the number of stable closed-loop poles

$-$ the number of stable open-loop poles.   (5.46)

(5.45) and (5.46) are equivalent because the number of unstable poles plus the number of stable poles is the same for the open-loop and for the closed-loop systems.

5.2.6 Open-loop poles on the $j\omega$-axis

To handle poles on the imaginary axis, one modifies the Nyquist contour slightly, in order to avoid the poles on the imaginary axis. This procedure is shown in Fig. 5.36. A half circle of radius ε is inserted in the path to avoid the poles. The rule $Z = N + P$ still applies, but P refers to the number of open-loop poles located in the modified contour, and Z refers to the number of closed-loop poles in the same area. Since the closed-loop poles differ from the open-loop poles (when there are no pole/zero cancellations in the open-loop transfer function), Z will still be equal to the number of unstable closed-loop poles if ε is sufficiently small. The only difficulty in the application of the modified Nyquist criterion comes from the need to transform the two small paths around the imaginary poles. The procedure is best explained through examples.

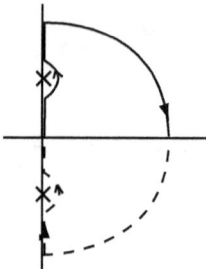

Figure 5.36: Modified Nyquist contour to handle poles on the $j\omega$-axis

Example 1: $G(s) = 1/\left(s(s+1)\right)$. The Bode plots and the Nyquist curve for this system are shown in Fig. 5.37. For the usual Nyquist contour, the transformed path grows to infinity as ω reaches the origin. In the modified contour, $G(j\varepsilon) = 1/j\varepsilon(1+j\varepsilon) \simeq 1/j\varepsilon = -j/\varepsilon$, where ε is an arbitrarily small number. On the negative side, $G(-j\varepsilon) \simeq -1/j\varepsilon = j/\varepsilon$. To count the encirclements, we need to connect the end of the branches at $\omega = \varepsilon$ and $\omega = -\varepsilon$.

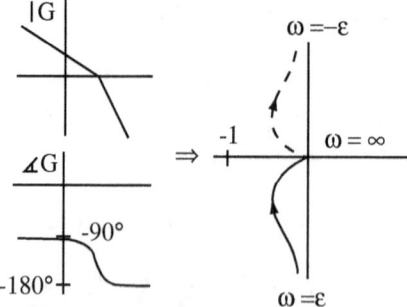

Figure 5.37: Nyquist curve for a system with a pole at the origin

Fig. 5.38 shows the detail of the modified contour around the origin. We assume that a half circle is used to connect the two paths, so that

$$s = \varepsilon \, e^{j\theta}, \text{ with } \theta = -\pi/2 \to \pi/2. \tag{5.47}$$

In the transformed path

$$G(s) = \frac{1}{s(s+1)} = \frac{1}{\varepsilon \, e^{j\theta}(1+\varepsilon \, e^{j\theta})} \simeq \frac{1}{\varepsilon \, e^{j\theta}} = \frac{1}{\varepsilon} e^{-j\theta}. \tag{5.48}$$

Since $\theta = -\pi/2 \to \pi/2$, the transformed path connects clockwise the branches at $90°$ and $-90°$. As a result, the transformed, modified contour may be connected as shown in Fig. 5.39. Since there are no unstable open-loop poles in the modified contour, $P = 0$. On the other hand, there are no encirclements of the $(-1,0)$ point by the transformed contour, so that $N = 0$. As a result, the closed-loop system is stable for all $k > 0$ (as predicted by the root-locus).

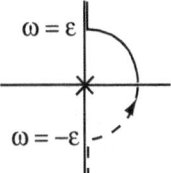

Figure 5.38: Detail of the modified contour in the vicinity of the origin

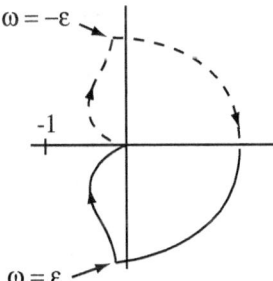

Figure 5.39: Nyquist diagram for $G(s) = 1/(s(s+1))$ and a modified Nyquist contour

Example 2: $G(s) = -1/(s(s+1))$. The Nyquist diagram is shown in Fig. 5.40. Since $N = 1$, $P = 0$, we have $Z = 1$, and there is one unstable closed-loop pole.

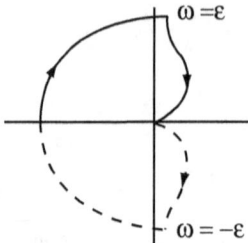

Figure 5.40: Nyquist diagram for $G(s) = -1/(s(s+1))$ and a modified Nyquist contour

Example 3: $G(s) = 1/(s^3(s+1))$. The Nyquist diagram is shown in Fig. 5.41. For $s = \varepsilon\, e^{j\theta}$

$$G(s) \simeq \frac{1}{\varepsilon^3}e^{-3j\theta}, \tag{5.49}$$

so that the transformed path sweeps from $270°$ to $-270°$. $P = 0$, $N = 2$, so that $Z = 2$ and there are two unstable closed-loop poles. This result may be verified from the root-locus shown in Fig. 5.42.

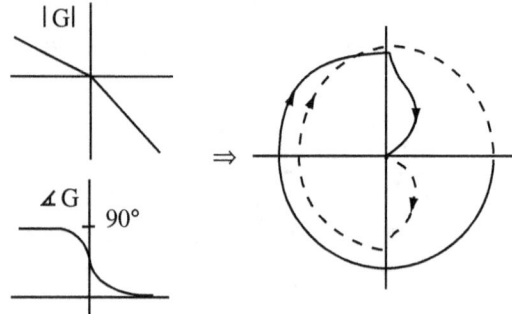

Figure 5.41: Bode plots (left) and Nyquist diagram (right) for $G(s) = 1/(s^3(s+1))$

General procedure

For a system with n poles at $s = 0$, the procedure can be applied in a similar manner. Plot $G(j\omega)$ for $\omega = \varepsilon \to \infty$, and draw $G(-j\omega)$ by symmetry. Then, connect $G(-j\varepsilon)$ to $G(j\varepsilon)$ with a circular curve that rotates around the origin by an angle $n \times 180°$ in the clockwise direction. Let P be the number of unstable

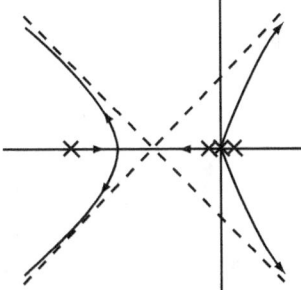

Figure 5.42: Root-locus for $G(s) = 1/(s^3(s+1))$

open-loop poles, not including the pole at $s = 0$, and count the number of encirclements N. The criterion is then applied as for the original contour. The same procedure may also be applied for pole(s) at $s = j\omega_0$, connecting the branches for $G(j\omega_0 - j\varepsilon)$ to $G(j\omega_0 + j\varepsilon)$.

5.3   Gain and phase margins

5.3.1   Gain margin

Aside from the concept of stability for a closed-loop system, an almost equally important consideration is how far the system is from instability. This is the concept behind gain and phase margins. Both apply to a system which is known to be closed-loop stable, but does not have to be open-loop stable.

By definition, the gain margin is the maximum value of the gain $k > 0$ by which the open-loop transfer function may be multiplied before the closed-loop system reaches instability. For example, consider the open-loop system

$$G(s) = \frac{1}{(s+1)^3},\qquad\qquad (5.50)$$

which was found to yield a stable closed-loop system. It was also found that the closed-loop system became unstable if the gain was multiplied by 8. Therefore, the gain margin of the system with open-loop transfer function $G(s)$ is equal to 8. Sometimes, the gain margin is expressed in dB. Then, $GM_{\mathrm{dB}} = 20\log(8) = 18$ dB.

5.3.2   Gain margin in the Nyquist diagram

In the Nyquist diagram, the gain margin is the value of the gain $k > 0$ by which the open-loop transfer function may be multiplied before the Nyquist curve passes through the $(-1, 0)$ point. Indeed, if the gain is increased so that $P(j\omega_1)C(j\omega_1) = -1$, for some frequency ω_1, then $s = \pm j\omega_1$ is a closed-loop pole and the closed-loop system is unstable. For larger values of the gain, the closed-loop system will not necessarily be unstable, but will almost always be so because the number of encirclements will be different. Fig. 5.43 shows how the gain margin is obtained from the Nyquist diagram. Assuming that the Nyquist curve crosses the real axis to the right of the $(-1, 0)$ point at $-a$, the gain margin is $1/a$.

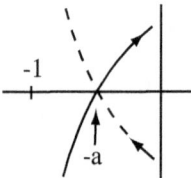

Figure 5.43: Computing the gain margin from the Nyquist diagram

For many systems, the gain may be reduced by any amount without resulting in instability. When one says that the gain margin is 2, it means that the gain may be multiplied by any number between 0 and 2. In some cases, however, the gain margin involves a lower number as well. The system is then called *conditionally stable* (see Figs. 5.30 and 5.31 involving a gain k restricted to being between 1 and 2). The nominal gain can neither be increased, nor decreased arbitrarily without resulting in instability. In the example shown in Fig. 5.44, the gain margin is given by

$$GM = (\frac{1}{b}, \frac{1}{a}). \tag{5.51}$$

5.3.3   Gain margin in the Bode plots

The computation of the gain margin in the Nyquist diagram may translated into the Bode plots. Let ω_1 be the frequency such that

$$\angle P(j\omega_1)C(j\omega_1) = 180° \pm n\,360°, \; n = 0, 1, 2, \cdots \tag{5.52}$$

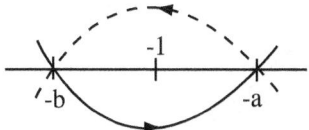

Figure 5.44: Gain margin for a conditionally stable system

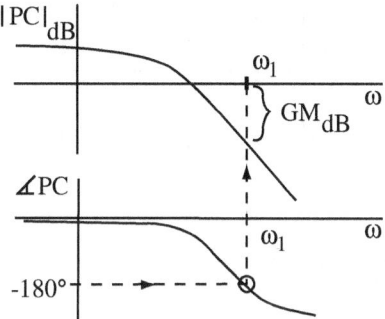

Figure 5.45: Determination of the gain margin from the Bode plots

Assuming $|P(j\omega_1)C(j\omega_1)| \leqslant 1$, the gain margin is given by

$$GM = \frac{1}{|P(j\omega_1)C(j\omega_1)|},$$ (5.53)

or

$$GM_{dB} = -20\log|P(j\omega_1)C(j\omega_1)|_{dB}.$$ (5.54)

The interpretation of the definition in Bode plots is shown in Fig. 5.45. If several frequencies are associated with an angle of 180°, the gain margin is the smallest value of all obtained.

If $|P(j\omega_1)C(j\omega_1)| > 1$ for one or more of the frequencies, the gain margin has a lower bound. Fig. 5.46 shows a hypothetical example where the gain margin is given by

$$GM_{dB} = (-6,\ 10) \Leftrightarrow GM = (\frac{1}{2},\ 3\ ).$$ (5.55)

Fig. 5.47 shows the associated Nyquist diagram.

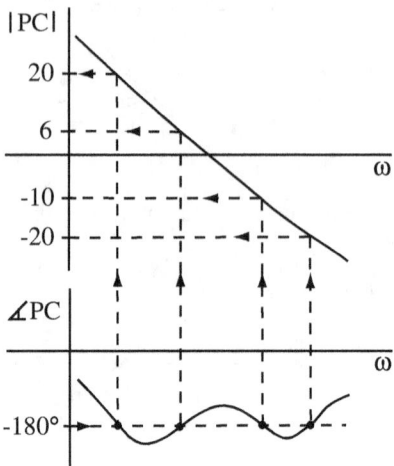

Figure 5.46: Bode plot with multiple intersections of the 180° line

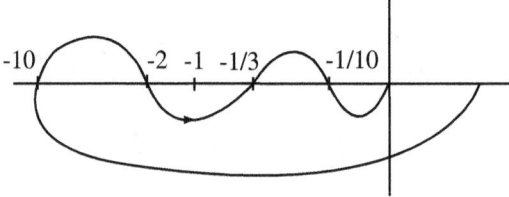

Figure 5.47: Nyquist diagram of a system with multiple intersections of the −180° line

5.3.4 Phase margin

The phase margin is the maximum angle that may be added to the phase of the frequency response before the closed-loop system becomes unstable. It is also the angle that must be added to the open-loop frequency response so that the Nyquist curve passes through the $(-1, 0)$ point. Because the frequency response of various actuators exhibits phase lags for sufficiently high frequencies, the phase margin quantifies the ability of the system to maintain stability despite such effects.

The phase margin may be determined from the Nyquist diagram as shown in Fig. 5.48. The circle of radius 1 is intersected with the Nyquist curve, and the angle between the negative real axis and the intersection point is the phase margin. If several points are found, the angle of smallest magnitude defines the phase margin. Because $P(-j\omega) = P^*(j\omega)$, the phase margin is the same in magnitude for both positive and negative directions.

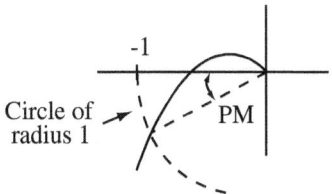

Figure 5.48: Determination of the phase margin in the Nyquist diagram

5.3.5 Phase margin in the Bode plots

To compute the phase margin in the Bode plots, one finds the frequency (or frequencies) ω_2 such that

$$|P(j\omega_2)C(j\omega_2)| = 1, \tag{5.56}$$

or

$$|P(j\omega_2)C(j\omega_2)|_{\mathrm{dB}} = 0. \tag{5.57}$$

Then, the phase margin is given by

$$PM = 180° - |\angle P(j\omega_2)C(j\omega_2)|. \tag{5.58}$$

where the angle function $\angle$ is defined between $-180°$ and $180°$. The concept is shown in Fig. 5.49. The frequency ω_2 is called the *crossover frequency*. If several frequencies are found, the smallest phase margin computed with the formula is the phase margin of the system.

5.3.6 Delay margin

There is a significant difference between the phase margin and the gain margin. In theory, the gain margin could be determined experimentally. However, a system with a frequency response $P(j\omega) = e^{j\varphi}$ for some phase φ positive or negative cannot be realized (the associated impulse response is not zero for

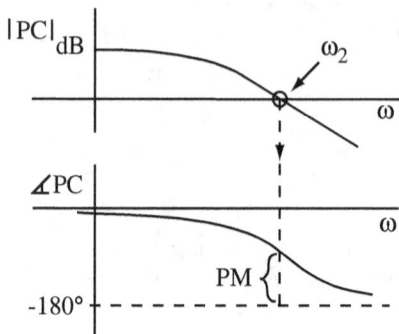

Figure 5.49: Determination of the phase margin from the Bode plots

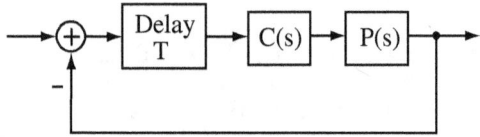

Figure 5.50: Feedback system for the definition of delay margin

$t < 0$). Therefore, the concept of phase margin can only be justified in terms of the Nyquist diagram.

A more practical concept is the *delay margin*, which is the maximum amount of time delay T that may be added to the open-loop transfer function before the closed-loop system becomes unstable. This concept is illustrated in Fig. 5.50. The delay margin can be estimated in practice by inserting a delay element in the loop (one would increase the delay until oscillations appear in the response of the system and before instability fully develops).

A delay T corresponds to a transfer function e^{-sT}, with

$$\left| e^{-j\omega T} \right| = 1,$$
$$\angle e^{-j\omega T} = -\omega T \text{ (in rad)}. \tag{5.59}$$

Therefore, considering the Nyquist criterion

$$\text{Phase margin (rad)} = \text{Crossover frequency (rad/s)}$$
$$\times \text{ Delay margin (s)}. \tag{5.60}$$

or

$$\begin{aligned} \text{Delay margin (s)} \quad &= \quad \frac{\text{Phase margin (rad)}}{\text{Crossover frequency (rad/s)}} \\ &= \quad \frac{\text{Phase margin (deg)}/360°}{\text{Crossover frequency (Hz)}}. \end{aligned} \qquad (5.61)$$

So, a phase margin of 45° with a crossover frequency of 1 kHz corresponds to a delay margin of 1/8 ms. If the crossover frequency was 10 Hz, the delay margin would be 100 times larger.

5.3.7 Relationship between phase margin and damping

There is a strong relationship between the phase margin and the damping of the poles of the closed-loop system as defined in (5.9). The relationship can be established analytically and exactly for the second-order system shown in Fig. 5.51. For $\zeta < 1$, the closed-loop poles are located at

$$s = -a \pm jb \quad \text{with } a = \zeta\omega_n \text{ and } b = \sqrt{1 - \zeta^2}\omega_n. \qquad (5.62)$$

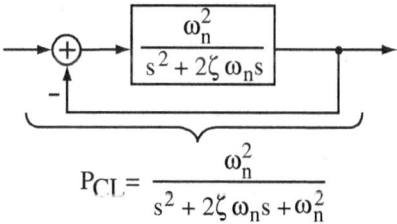

$$P_{CL} = \frac{\omega_n^2}{s^2 + 2\zeta\,\omega_n s + \omega_n^2}$$

Figure 5.51: Second-order system to relate damping and phase margin

The following results may be derived analytically for this system.

1. **Phase margin (5.58)**

$$PM = \tan^{-1}\left(\frac{2\zeta}{\left((4\zeta^4 + 1)^{1/2} - 2\zeta^2\right)^{1/2}}\right) \quad \text{(rad)}$$

$$\simeq 100\zeta \quad \text{(deg)}. \qquad (5.63)$$

2. **Percent overshoot in the step response (3.45)**

$$PO = e^{-\zeta\pi/\sqrt{1-\zeta^2}} \times 100 \ \ (\%)$$
$$\simeq e^{-\zeta\pi} \times 100 \ \ (\%). \tag{5.64}$$

3. **Peaking of the frequency response (5.13)**

$$PF = \max_\omega \frac{|P_{CL}(j\omega)|}{|P_{CL}(0)|} = \frac{1}{2\zeta} \frac{1}{\sqrt{1-\zeta^2}} \simeq \frac{1}{2\zeta}. \tag{5.65}$$

With these relationships, the following table can be derived.

ζ	PM	PO	PF
0.2	22.6°	52.7%	2.55
0.3	33.3°	37.2%	1.75
0.4	43.1°	25.4%	1.36
0.5	51.8°	16.3%	1.15
0.6	59.2°	9.5%	1.04
0.7	65.2°	4.6%	1.0002

The results show a tight connection between the phase margin of the second-order system, the overshoot of the step response, and the peaking of the frequency response. In view of the results, a phase margin of 60° is often taken as an objective in control systems. Although the formulas for the phase margin were obtained for a specific second-order system, the results are taken to provide guidance for higher-order systems as well.

5.3.8 Frequency-domain design

Although the root-locus technique is helpful for the design of feedback systems, frequency-domain design is also effective is also effective, with definite advantages in some situations.

Consider the feedback system shown in Fig. 5.52, where $G(s)$ is the combined transfer function of the plant and compensator. The closed-loop transfer function and the closed-loop frequency response are given by

$$G_{CL}(s) = \frac{G(s)}{1+G(s)}, \quad G_{CL}(j\omega) = \frac{G(j\omega)}{1+G(j\omega)}. \tag{5.66}$$

The objective of tracking translates into an objective that $G_{CL}(j\omega) \simeq 1$, which may be achieved by setting $|G(j\omega)| \gg 1$. Although it would be desirable to

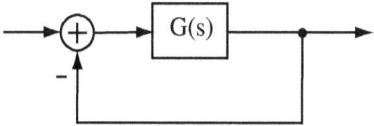

Figure 5.52: Feedback system

have this property hold true for all ω, it is only practical to do so for a finite range of frequencies. Indeed, the gain of physical systems usually falls rapidly at high frequencies. A controller may only partially compensate for this effect. The phase of physical systems also tends to increase rapidly at high frequencies, and often does so in ways that cannot be precisely modelled. In order to ensure closed-loop stability, it may be necessary to bring the loop gain well below 1 before significant phase variations are observed.

Overall, the design problem in the frequency domain consists first of all in a careful selection of the crossover frequency of the closed-loop system. Below this frequency, the loop gain will be made as large as possible (sometimes through the use of integral control). Around the crossover frequency, the phase will be carefully controlled in order to ensure stability as well as robustness to uncertainties and parameter variations (considering the Nyquist criterion). A Bode plot of a typical open-loop transfer function is shown in Fig. 5.53.

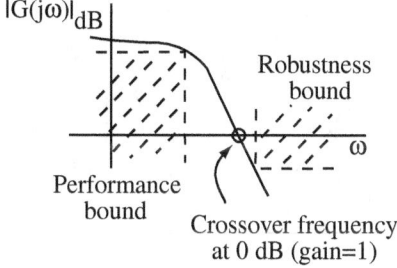

Figure 5.53: Design objectives on the loop frequency response

5.3.9 Example of frequency-domain design with a lead controller

Lead controller: consider the so-called *lead controller*

$$C(s) = k_c \frac{(s+b)}{(s+a)},$$ (5.67)

where $a > b > 0$. In the s–plane, the controller has a pole and a zero on the negative real axis, with the zero being closer to the origin than the pole.

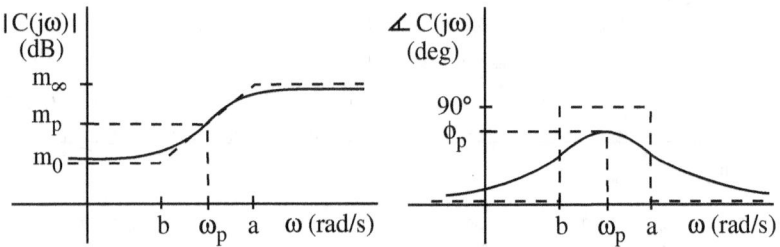

Figure 5.54: Bode plots of lead controller

The Bode plots of the lead controller are shown in Fig. 5.54. The controller is called a lead controller because the phase angle is positive. For the same reason, the controller is called a *lag controller* if $b > a$ (the system becomes a proportional-integral controller for $a = 0$). The DC gain m_0 and the high-frequency gain m_∞ of the controller are

$$m_0 = k_c \frac{b}{a}, \qquad m_\infty = k_c.$$ (5.68)

The frequency ω_p is defined on the figure as the frequency where the angle of the frequency response is maximum. ϕ_p is the phase of the lead controller at the frequency ω_p. Since

$$\begin{aligned}
|C(j\omega)| &= k_c \sqrt{\frac{b^2 + \omega^2}{a^2 + \omega^2}}, \\
\angle C(j\omega) &= \tan^{-1}(\frac{\omega}{b}) - \tan^{-1}(\frac{\omega}{a}), \\
\frac{d\angle C(j\omega)}{d\omega} &= \frac{1}{1 + \omega^2/b^2} \frac{1}{b} - \frac{1}{1 + \omega^2/a^2} \frac{1}{a} = \frac{(\omega^2 - ab)(b - a)}{(b^2 + \omega^2)(a^2 + \omega^2)},
\end{aligned}$$ (5.69)

one can deduce that

$$\omega_p = \sqrt{ab},$$

$$m_p = k_c\sqrt{\frac{b}{a}},$$

$$\phi_p = \tan^{-1}\left(\sqrt{\frac{a}{b}}\right) - \tan^{-1}\left(\sqrt{\frac{b}{a}}\right). \tag{5.70}$$

The result shows that the frequency of the peak is located mid-way between the pole and the zero on a log scale $(\log(\omega_p) = (\log(a) + \log(b))/2)$.

A different formula can be obtained for ϕ_p, using the fact that

$$C(j\omega) = k_c\frac{(b+j\omega)(a-j\omega)}{(a^2+\omega^2)} = k_c\frac{(ab+\omega^2)+j\omega(a-b)}{(a^2+\omega^2)}, \tag{5.71}$$

so that

$$\sin^2(\angle C(j\omega_p)) = \sin^2(\phi_p) = \frac{\omega_p^2(a-b)^2}{(ab+\omega_p^2)^2+\omega_p^2(a-b)^2} = \frac{(a-b)^2}{(a+b)^2}, \tag{5.72}$$

and

$$\sin(\phi_p) = \frac{(a/b-1)}{(a/b+1)} \Leftrightarrow \frac{a}{b} = \frac{1+\sin(\phi_p)}{1-\sin(\phi_p)}. \tag{5.73}$$

This result shows that the amount of phase depends on the ratio of the pole magnitude over the zero magnitude (a/b). Specifically

a/b	ϕ_p
5.83	45°
9	53.1°
13.9	60°

Lead controller for a double integrator: consider now the control problem for a double integrator

$$P(s) = \frac{k}{s^2}. \tag{5.74}$$

The Bode plots of the loop transfer function are shown in Fig. 5.55. From root-locus theory, the closed-loop system is known to be stable for all $kk_c > 0$. A possible choice is to set the gain of the controller such that the loop gain is equal to 1 at ω_p. In this manner, the frequency ω_p is the same as the crossover frequency of the system, and the phase ϕ_p is the phase margin. For the plant under consideration, setting the loop gain to be 1 at ω_p implies that

$$m_p \frac{k}{\omega_p^2} = k_c \sqrt{\frac{b}{a}} \frac{k}{\omega_p^2} = 1, \tag{5.75}$$

or

$$k_c = \sqrt{\frac{a}{b}} \frac{\omega_p^2}{k}. \tag{5.76}$$

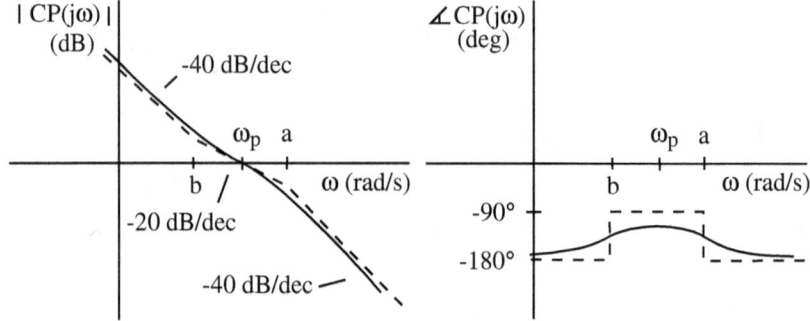

Figure 5.55: Bode plots of lead controller with double integrator plant

Assume that a certain phase margin and a certain cross-over frequency are specified. The ratio a/b is determined by the phase margin and ω_p is set equal to the crossover frequency. Then, the controller parameters are determined from

$$a = \omega_p \sqrt{\frac{a}{b}}, \quad b = \omega_p \sqrt{\frac{b}{a}}, \quad k_c = \sqrt{\frac{a}{b}} \frac{\omega_p^2}{k}. \tag{5.77}$$

For example, if a phase margin of 53.1° is chosen, $a/b = 9$, and the parameters of the controller are equal to

$$a = 3\omega_p, \quad b = \frac{\omega_p}{3}, \quad k_c = \frac{3\omega_p^2}{k}. \tag{5.78}$$

Therefore, the three parameters of the controller can be set as functions of the crossover frequency ω_p. ω_p is a free parameter that can be adjusted experimentally to be as large as possible to maximize the bandwidth of the system. In theory, there is no limit to ω_p, but in practice, additional dynamics will limit the possible range. If these dynamics were included in the model, a careful design would maximize the closed-loop bandwidth within gain and phase margin specifications. Such design sinvolve tedious trial-and-error adjustments of the controller parameters, and is preferably performed nowadays using some numerical optimization tool [9].

5.3.10   Design in the Nyquist diagram

The gain and phase margins aim at quantifying the ability of a feedback system to tolerate uncertainties or variations in the nominal plant transfer function. This property is generally referred to as the *robustness* of the feedback system. The gain and phase margins are convenient measures of robustness, but some pathological cases may be conceived, such as the one shown in the Nyquist diagram of Fig. 5.56. For this system, $GM = \infty$, $PM > 45°$. However, small but *combined* changes in gain and phase could lead to instability. Generally, the number $|1 + G(j\omega)|$ represents the distance from the Nyquist curve to the $(-1, 0)$ point, and is a good measure of robustness, although its interpretation is not as intuitive as the gain and phase margins.

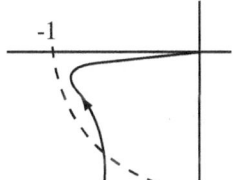

Figure 5.56: Nyquist diagram of a non-robust design with good gain and phase margins

An example of a design in the Nyquist diagram is shown in Fig. 5.57. For a range of frequencies where tracking is required, the loop gain is made large. Around the crossover frequency, the loop frequency response is carefully controlled so that a sufficient phase margin is guaranteed and is insensitive to variations of the gain of the system. When the loop gain becomes sufficiently small at high frequencies, the phase behavior becomes irrelevant.

Peaking of the closed-loop frequency response should also be avoided, since large peaks translate into poor transient responses, and sensitivity to noise and disturbances at the associated frequencies. Fig. 5.58 shows $|G_{CL}(j\omega)|$ as a function of $\text{Re}(G(j\omega))$ and $\text{Im}(G(j\omega))$. As expected, the gain is infinite at the $(-1, 0)$ point, and is large in its vicinity.

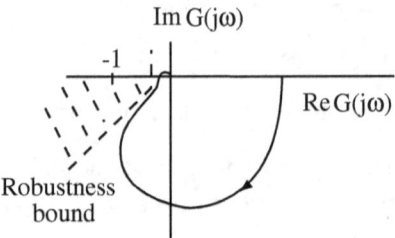

Figure 5.57: Robustness objective in the Nyquist diagram

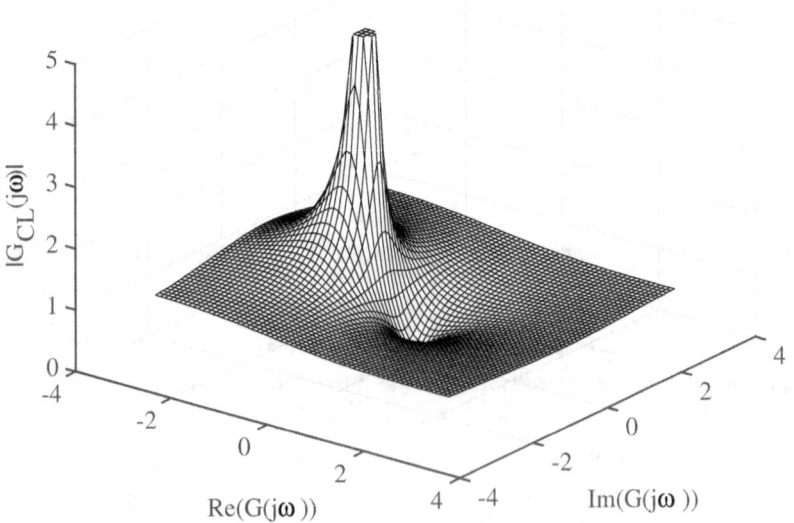

Figure 5.58: Plot of $|G_{CL}(j\omega)|$ as a function of $\text{Re}(G(j\omega))$ and $\text{Im}(G(j\omega))$

The plot is often represented through level lines, as shown in Fig. 5.59. It turns out that $|G_{CL}(j\omega)| = M$ if $G(j\omega)$ belongs to a circle of radius $M/(M^2 - 1)$ with center $-M^2/(M^2 - 1)$. In order to avoid peaking in the frequency domain, the Nyquist curve of the open-loop frequency response should avoid as much as possible the portion of the complex plane where $\mathrm{Re}(G(j\omega)) < -1/2$, especially when $\mathrm{Im}(G(j\omega))$ is small.

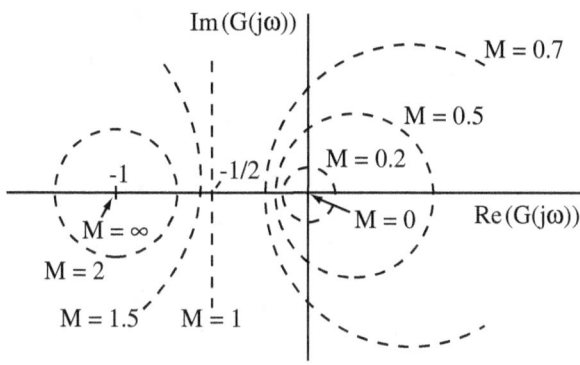

Figure 5.59: Lines of constant closed-loop magnitude

5.4 Problems

Problem 5.1: Sketch the Bode plots for the following transfer functions. Make sure to label the graphs, and to give the slopes of the lines in the magnitude plot.

(a) $P(s) = \dfrac{s - 10}{(s + 1)(s + 100)}$

(b) $P(s) = \dfrac{100(s - 10)(s + 10)}{(s + 0.1)(s + 100)^2}$

(c) $P(s) = \dfrac{(s + 10)}{s^2 + 0.1s + 1}$

(d) $P(s) = \dfrac{s - 10}{s(s + 1)}$

(e) $P(s) = \dfrac{100}{(s - 10)^2(s + 1)}$

(f) $P(s) = \dfrac{s^2 + 2s + 100}{s^2}$

Problem 5.2: (a) The magnitude Bode plot of a system is shown in Fig. 5.60. What are the possible transfer functions of *stable* systems having this Bode plot?

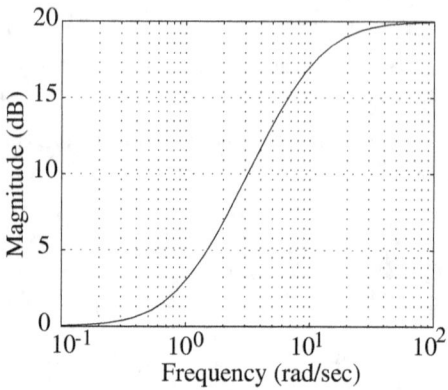

Figure 5.60: Bode plot for problem 5.2 (a)

(b) The Bode plots of a system are shown in Fig. 5.61. Give an estimate of the transfer function of the system.

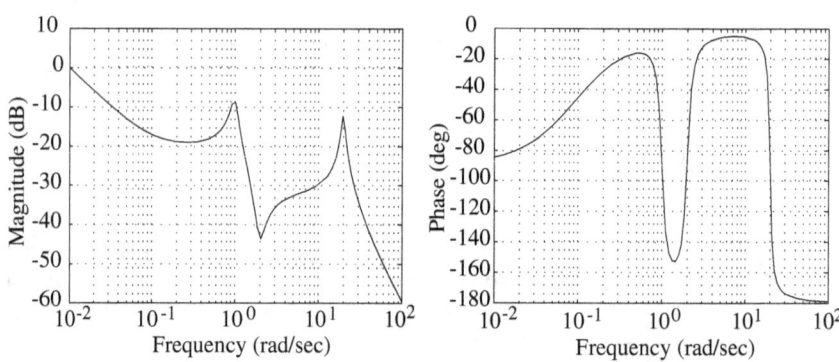

Figure 5.61: Bode plots for problem 5.2 (b)

Problem 5.3: (a) The system whose Bode plots are shown in Fig. 5.62 is stable in closed-loop. Find its gain margin and phase margin.

(b) Describe the behavior of the closed-loop system of part (a) if the open-loop gain is increased to a value close to the maximum value given by the gain margin. In particular, what can you say about the locations of the poles of the closed-loop system?

(c) Consider an open-loop stable system such that the magnitude of its frequency

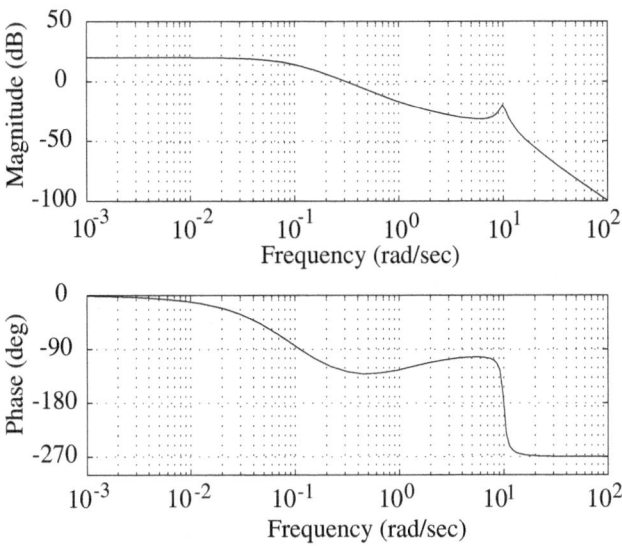

Figure 5.62: Bode plots for problem 5.3

response is less than 1 for all ω. Can it be determined whether the closed-loop system is stable with only that information?

Problem 5.4: (a) The Nyquist diagram of a *stable* system is shown in Fig. 5.63, with the overall diagram shown on the left and the detail around the (-1,0) point shown on the right. The solid line corresponds to $\omega > 0$, with the arrow giving the direction of increasing ω. The dashed line is the symmetric curve obtained for $\omega < 0$. Assuming that the transfer function of the system is multiplied by a gain $k > 0$, what is the set of values of k for which the system is stable in closed-loop ?

(b) Repeat part (a) for the system whose Nyquist curve is shown in Fig. 5.64, given that the system has *one* unstable pole.

Problem 5.5: (a) The Nyquist diagram for $P(s) = 5(s + 2)/(s + 1)^3$ is shown in Fig. 5.65, with the overall diagram shown on the left and the detail around the (-1,0) point shown on the right. Indicate what the gain margin and the phase margin are (for the phase margin, a rough estimate is fine). Compare the gain margin results with those predicted by a root-locus plot and the use of the Routh-Hurwitz criterion.

(b) Repeat part (a) for $P(s) = 2(s + 5)/(s + 1)^3$ and the diagrams shown in

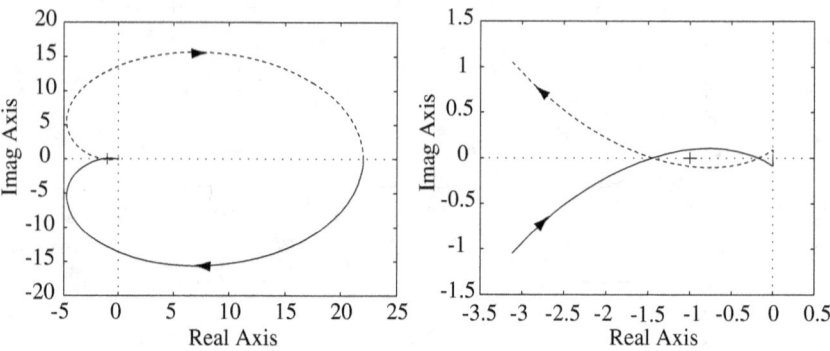

Figure 5.63: Nyquist diagram for problem 5.4 (a)

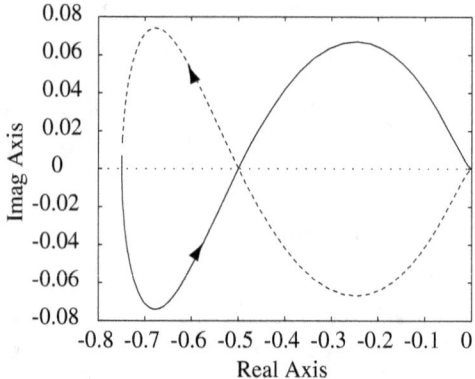

Figure 5.64: Nyquist curve for problem 5.4 (b)

Fig. 5.66.

Problem 5.6: Sketch the Bode plots for the following transfer function

$$P(s) = \frac{10(s-1)}{(s+10)^2}.$$ (5.79)

Make sure to label the graphs, and to give the slopes of the lines in the magnitude plot.

Problem 5.7: Sketch the Bode plots for the following transfer function

$$P(s) = \frac{10(s+1)}{s^2(s^2 - 2s + 100)}.$$ (5.80)

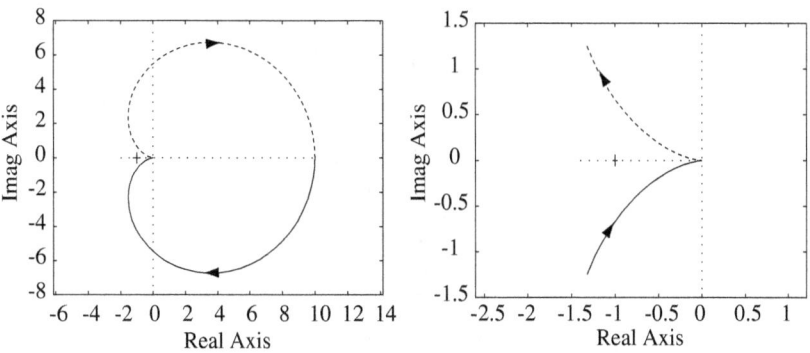

Figure 5.65: Nyquist curve for problem 5.5 (a)

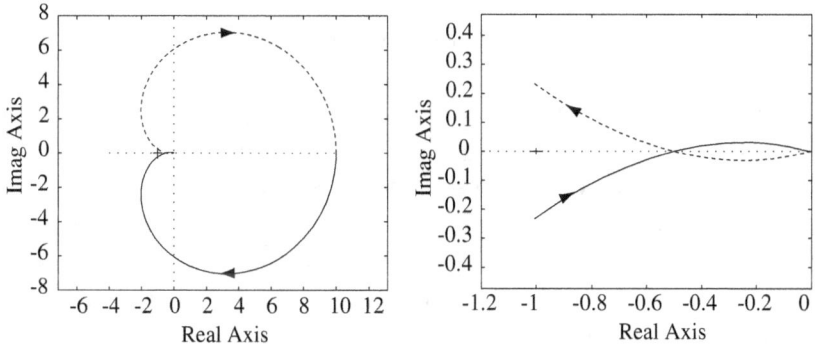

Figure 5.66: Nyquist curve for problem 5.5 (b)

Make sure to label the graphs, and to give the slopes of the lines in the magnitude plot.

Problem 5.8: (a) Sketch the Bode plots for

$$P(s) = \frac{(s-1)}{s(s+1)}. \tag{5.81}$$

Be sure to label the axes precisely.

(b) Sketch the Bode plots for a transfer function whose poles are placed in the s−plane as shown on Fig. 5.67. Assume that the DC gain of the system is 10 (with a positive sign). There are five poles located on a circle of radius 10. One pole is real, two poles are on a $45°$ line, and two poles have real parts equal to -0.5. You may use the fact that, for α small, $\sin(\alpha) \simeq \tan(\alpha) \simeq \alpha$, and $\cos(\alpha) \simeq 1$. Be sure to label the axes precisely.

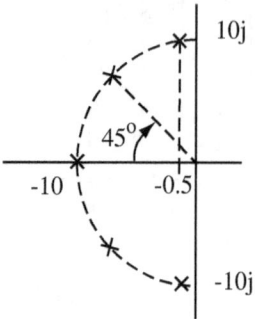

Figure 5.67: Pole locations for problem 5.8 (b)

Problem 5.9: (a) Give the gain margin and the phase margin of the system whose Bode plots are shown in Fig. 5.68 (the plots are for the open-loop transfer function and the closed-loop transfer function is assumed to be stable).

(b) Indicate whether the system whose Nyquist curve is shown in Fig. 5.69 is closed-loop stable, given that it is open-loop stable.

(c) What are the values of the gain $g > 0$ by which the open-loop transfer function of part (b) may be multiplied, with the closed-loop system remaining stable?

(d) Sketch an example of a Nyquist curve for a system that has three unstable open-loop poles, and which is closed-loop stable.

Problem 5.10: The magnitude Bode plot of a system is shown in Fig. 5.70. Give all the possible transfer functions of systems having this Bode plot. The poles and zeros are all real, and the values of the gain, poles, and zeros, are all multiples of 10.

Problem 5.11: All parts of this problem refer to the system whose Nyquist curve is shown in Fig. 5.71 (only the portion for $\omega > 0$ is plotted).

(a) Knowing that the *closed-loop* system is stable, can one say for sure that the *open-loop* system is stable?

(b) Given that the closed-loop system is stable, estimate the gain margin and the phase margin of the closed-loop system.

(c) How many unstable poles does the closed-loop system have if the open-loop

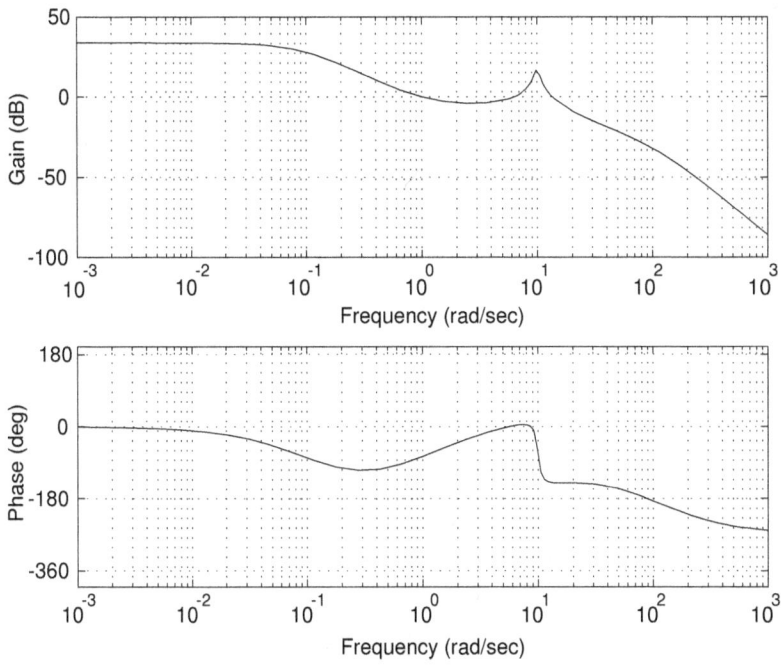

Figure 5.68: Bode plots for problem 5.9 (a)

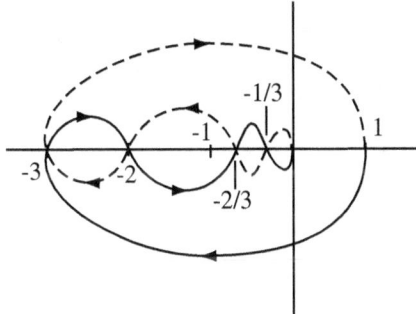

Figure 5.69: Nyquist curve for problem 5.9 (b)

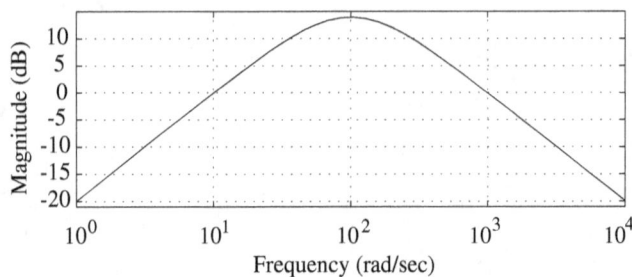

Figure 5.70: Bode plot for problem 5.10

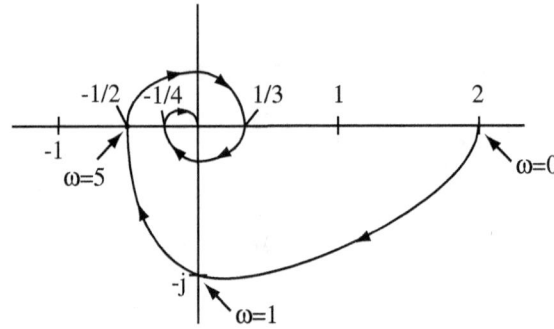

Figure 5.71: Nyquist curve for problem 5.11

gain is multiplied by 5?

(d) Give the steady-state response $y_{ss}(t)$ of the *open-loop* system to an input $x(t) = 2$. Repeat for $x(t) = 3\cos(t)$ and for $x(t) = 4\cos(5t)$.

(e) Repeat part (d) for the *closed-loop* system.

Problem 5.12: (a) Sketch the Bode plots of

$$G(s) = \frac{1}{(s+1)(s-1)}. \tag{5.82}$$

Be sure to label the axes precisely.

(b) Sketch the *magnitude* Bode plot of

$$G(s) = \frac{(s^2 + 1)(s - 100)}{s(s + 10)^2}. \tag{5.83}$$

Only the magnitude is needed. Be sure to label the axes precisely.

Problem 5.13: (a) Consider the Nyquist diagram of a transfer function $G(s)$ shown in Fig. 5.72. Only the portion for $\omega > 0$ is plotted. Assume that $G(s)$ has no poles in the open right half-plane, and that a gain k is cascaded with $G(s)$. Find the ranges of positive k for which the closed-loop system is stable.

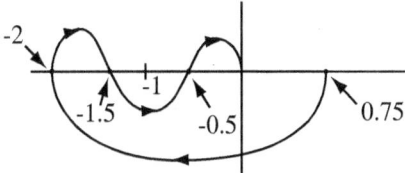

Figure 5.72: Nyquist curve for problem 5.13 (a)

(b) Bode plots of the open-loop transfer function of a feedback system are shown in Fig. 5.73, with the detail from 1 to 10 rad/s shown on the right. For this system:

- How much can the open-loop gain be changed (increased and/or decreased) before the closed-loop system becomes unstable ?

- What is a rough estimate of the phase margin of the feedback system?

The numerical results do not have to be precise.

(c) For the system of part (b), give the steady-state response of the open-loop system *and* of the closed-loop system to an input $x(t) = 2$.

Problem 5.14: (a) Consider the lead controller for the double integrator. For the design that makes the crossover frequency equal to ω_p, obtain the polynomial that specifies the closed-loop poles (as a function of a/b and ω_c). Show that one closed-loop pole is at $s = -\omega_c$ no matter what a/b is.

(b) Compute the other closed-loop poles as functions of ω_c, when $a/b = 5.83$, 9, and 13.9.

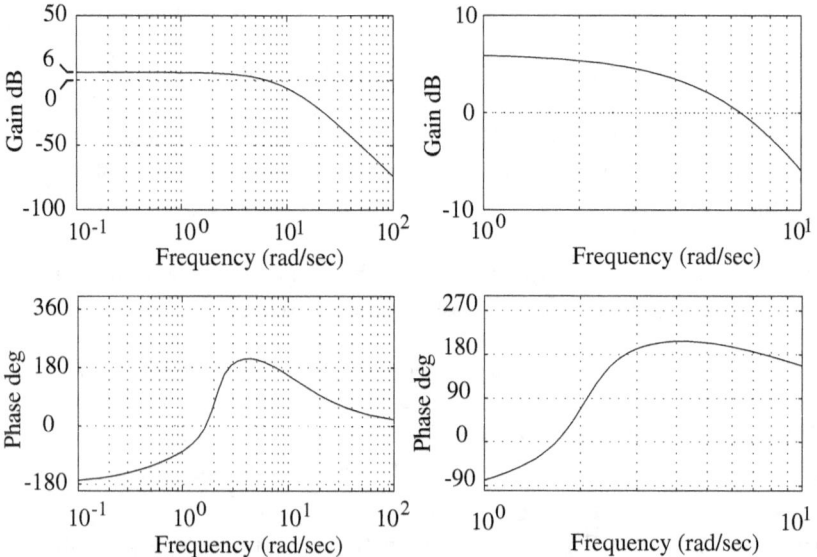

Figure 5.73: Bode plots for problem 5.13 (b)

Chapter 6

Discrete-time signals and systems

6.1 The z-transform

6.1.1 Discrete-time signals

A *discrete-time signal* is a function of time $x(k)$, where time is defined over a set of integer values, or $k = 0, 1, 2, \cdots$. A discrete-time signal is similar to a *sequence*, as defined in mathematical analysis. The discrete time values are sometimes called *steps,* or *samples.* Fig. 6.1 shows an example of a discrete-time signal.

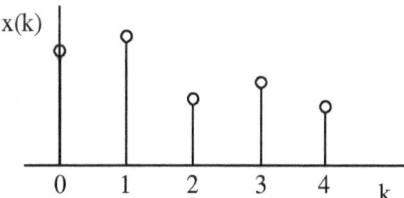

Figure 6.1: Discrete-time signal

In mathematical software packages, discrete-time signals are represented in various manners. Fig. 6.2 shows two standard options. The plot on the left is usually preferred, but note that the signal is only defined at the discrete-time instants, although the values between time instants are interpolated. For the plot on the right, options also include dots, + signs, and other symbols.

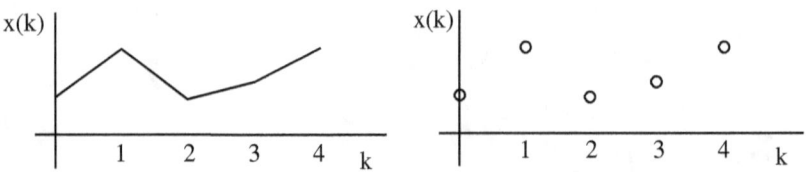

Figure 6.2: Two representations of a discrete-time signal

6.1.2 The z-transform

The z-transform of a signal $x(k)$ is defined as the function $X(z)$ such that

$$
\begin{aligned}
X(z) &= \sum_{k=0}^{\infty} x(k) z^{-k} \\
&= x(0) + x(1)z^{-1} + x(2)z^{-2} + \cdots
\end{aligned}
\tag{6.1}
$$

The z-transform is the result of an *infinite series*, involving all the values of the signal $x(k)$ and the complex variable z. The variable z is similar to the variable s of the Laplace transform. The definition of the z-transform is *unilateral, i.e.,* independent of $x(k)$ for $k < 0$. The *bilateral* z-transform requires summation over both positive and negative values of time, but is not used here. We discuss a few important examples.

1. **Discrete-time impulse**

   A discrete-time impulse is defined by

   $$
   x(k) = \delta(k) \Leftrightarrow x(0) = 1 \text{ and } x(k) = 0 \text{ for } k \neq 0.
   \tag{6.2}
   $$

   The signal is shown in Fig. 6.3. Its transform is easily found to be

   $$
   x(k) = \delta(k) \Leftrightarrow X(z) = 1.
   \tag{6.3}
   $$

   The result is the same as in continuous-time, but there is a considerable difference in that the time function is an ordinary function, rather than a generalized function (or distribution, or delta function) in continuous-time.

2. **Finite-length signal**

   A finite-length signal is a signal that vanishes after a finite period of time, *i.e.,*

   $$
   x(k) = 0 \quad \text{for} \quad k > N.
   \tag{6.4}
   $$

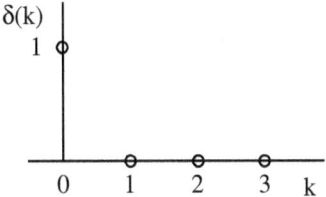

Figure 6.3: Discrete-time impulse

The transform of such a signal is given by

$$
\begin{aligned}
X(z) &= x(0) + x(1)z^{-1} + \cdots x(N)z^{-N} \\
&= \frac{x(0)z^N + x(1)z^{N-1} + \cdots + x(N)}{z^N}.
\end{aligned}
\tag{6.5}
$$

The z-transform of a finite-length signal is a rational function of z whose poles are all located at $z = 0$. Conversely, any rational function of z with all poles at $z = 0$ is the transform of a finite-length signal, and the signal is easily obtained from $X(z)$. For example

$$
X(z) = \frac{z^3 - z^2 + 2}{z^5} \qquad \Leftrightarrow \qquad x(k) = 0,\ 0,\ 1,\ -1,\ 0,\ 2,\ 0,\ 0,\ \dots
\tag{6.6}
$$

Note that the definition of the z-transform implies that a rational transform $X(z) = N(z)/D(z)$ must always be such that $\deg(N(z)) \leqslant \deg(D(z))$ (i.e., the transform must be a *proper* function of z)

3. **Step signal**

   A step signal is given by

$$
\begin{aligned}
x(k) &= 1 \quad \text{for } k \geqslant 0 \\
&= 0 \quad \text{for } k < 0.
\end{aligned}
\tag{6.7}
$$

   The transform is given by

$$
X(z) = 1 + z^{-1} + z^{-2} + \cdots
\tag{6.8}
$$

   The infinite series has an analytic expression

$$
X(z) = \frac{1}{1 - z^{-1}} = \frac{z}{z - 1}.
\tag{6.9}
$$

(6.9) may be obtained through the following auxiliary result. Since

$$
\begin{aligned}
\left(1 + a + a^2 + \cdots a^n\right)(1 - a) &= 1 + a + a^2 + \cdots + a^n \\
&\quad -a - a^2 - \cdots - a^{n+1} \\
&= 1 - a^{n+1}. \tag{6.10}
\end{aligned}
$$

It follows that

$$
1 + a + a^2 + \cdots a^n = \frac{1 - a^{n+1}}{1 - a} \tag{6.11}
$$

and

$$
\lim_{n \to \infty} \sum_{i=0}^{n} a^n = \frac{1}{1 - a} \qquad \text{if } |a| < 1. \tag{6.12}
$$

Applying the auxiliary result to the step signal, one finds that

$$
X(z) = \frac{1}{1 - z^{-1}} = \frac{z}{z - 1} \qquad \text{if } |z^{-1}| < 1, \text{ or } |z| > 1. \tag{6.13}
$$

As for the Laplace transform, the z-transform is defined only in the region of the z-plane where convergence of the infinite series is guaranteed. This region is called the *region of convergence (ROC)*. For the step signal, the region of convergence is the portion of the z plane located outside the circle of radius 1.

The z-transform of the step signal (6.9) is a rational function of z. The pole is at $z = 1$ and there is a zero at $z = 0$. The transform is different from the transform of the continuous-time step function, which is $(1/s)$. We will find that $z = 1$ is the equivalent of $s = 0$ in the s−plane.

4. **Geometric progression**

The geometric progression signal is the equivalent of the exponential signal in continuous-time, and is given by

$$
x(k) = a^k. \tag{6.14}
$$

For the time being, we assume that a is real. Depending on the magnitude of a, the signal has the following properties

$$
\begin{aligned}
|a| &< 1 \qquad \text{a decaying exponential} \\
|a| &> 1 \qquad \text{a growing exponential} \\
|a| &= 1 \qquad \text{a step signal}
\end{aligned}
$$

The z-transform is given by

$$X(z) = 1 + az^{-1} + a^2z^{-2} \cdots$$
$$= \frac{1}{1 - az^{-1}} = \frac{z}{z - a} \quad \text{if } |az^{-1}| < 1, \text{ or } |z| > a. \quad (6.15)$$

Fig. 6.4 shows the decaying signal for $|a| < 1$, and the associated pole in the z-plane.

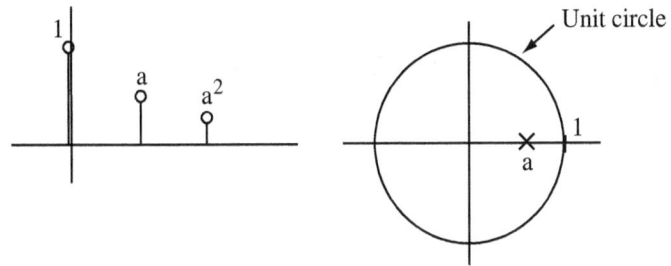

Figure 6.4: Exponentially decaying signal (geometric progression)

For the geometric progression, the region of convergence (ROC) is the portion of the z-plane located outside the circle of radius $z = a$. This feature is shown in Fig. 6.5. For rational transforms, the ROC is an open set whose boundary is the smallest circle that includes all the poles. The transforms of signals with different ROC's are defined in a common region sufficiently far from the origin, and one does not need to worry about the ROC in order to apply the transform to combinations of signals (this result only applies to the unilateral transform).

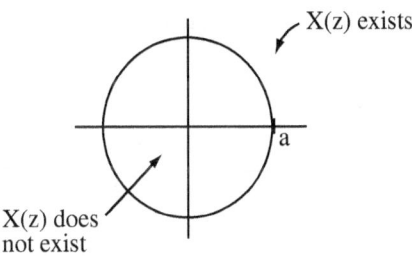

Figure 6.5: Region of convergence of the z-transform

5. **Signals with two complex poles**

The property that

$$x(k) = a^k \Leftrightarrow X(z) = \frac{z}{z-a} \qquad (6.16)$$

also applies to complex signals. As for the Laplace transform, we consider the signal

$$x(k) = cp^k + c^*p^{*k} \Leftrightarrow X(z) = \frac{cz}{z-p} + \frac{c^*z}{z-p^*}. \qquad (6.17)$$

As in Fig. 6.6, one defines the magnitude and angle of the complex pole with

$$p = |p|e^{j\angle p} \qquad -\pi < \angle p \leqslant \pi. \qquad (6.18)$$

Then, the time signal is

$$\begin{aligned} x(k) &= 2 \ \mathrm{Re}(cp^k) = 2|cp^k| \cos\left(\angle(cp^k)\right) \\ &= 2|c| \ |p|^k \cos(\angle p \ k + \angle c). \end{aligned} \qquad (6.19)$$

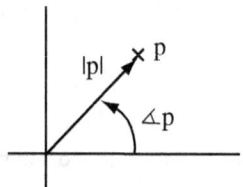

Figure 6.6: Magnitude/angle representation of a pole

The discrete-time result

$$x(k) = 2|c| \ |p|^k \cos(\angle p \ k + \angle c) \Leftrightarrow X(z) = \frac{cz}{z-p} + \frac{c^*z^*}{z-p^*} \qquad (6.20)$$

can be compared to the continuous-time result

$$x(t) = 2|c| \ e^{\mathrm{Re}(p)t} \cos(\mathrm{Im}(p) \ t + \angle c) \Leftrightarrow X(s) = \frac{c}{s-p} + \frac{c^*}{s-p^*}. \qquad (6.21)$$

Both signals consist in the product of an exponential signal and a sinusoidal signal. In discrete-time, however, the magnitude and angle of the pole play the role of the real part and imaginary part of the pole in continuous-time. Specifically,

$|p|$ determines the rate of growth/decay of the signal
  If $|p| > 1$, the signal grows
  If $|p| < 1$, the signal decays
  If $|p| = 1$, the signal is a pure sinusoid
$\angle p$ determines the frequency of the signal (see next section)

6. **Sinusoidal signals**

A sinusoidal signal is obtained as the special case of (6.20) with $|p| = 1$. For example, $x(k) = \cos(\Omega_0 k)$ corresponds to $|c| = 1/2$, $|p| = 1$, $\angle p = \Omega_0$, and $\angle c = 0$. Therefore,

$$p = e^{j\Omega_0} = \cos(\Omega_0) + j\sin(\Omega_0), \quad c = 1/2, \tag{6.22}$$

and the transform is

$$
\begin{aligned}
x(k) \;&=\; \cos(\Omega_0 k) \Leftrightarrow X(z) = \frac{1}{2}\frac{z}{z - e^{j\Omega_0}} + \frac{1}{2}\frac{z}{z - e^{-j\Omega_0}} \\
&=\; \frac{z^2 - \cos(\Omega_0)z}{z^2 - 2\cos(\Omega_0)z + 1}.
\end{aligned}
\tag{6.23}
$$

Similarly, the transform of a sin function can be found to be

$$
\begin{aligned}
x(k) = \sin(\Omega_0 k) \Leftrightarrow X(z) &= \frac{1}{2j}\frac{z}{z - e^{j\Omega_0}} - \frac{1}{2j}\frac{z}{z - e^{-j\Omega_0}} \\
&= \frac{\sin(\Omega_0)z}{z^2 - 2\cos(\Omega_0)z + 1}.
\end{aligned}
\tag{6.24}
$$

The two poles of the sinusoidal signals are on the *unit circle, i.e.,* the circle such that $|p| = 1$. This is shown in Fig. 6.7.

The angle between the real axis and the poles is the *discrete-time frequency* Ω_0, expressed in radians, or radians per sample. The period of the signal is $T = 2\pi/\Omega_0$. When T is an integer, the signal is periodic, and T is the number of samples associated with the period. Fig. 6.8 shows a few sinusoidal signals with integer periods. When T is a rational number, with $T = n/m$ and n, m integers, the signal is sinusoidal with period n. Indeed, $\cos(2\pi(m/n)k)$ repeats itself when n is added to k. When T is not an integer or a rational number, the signal is not periodic over an integer number of samples, but still exhibits the same sinusoidal shape. As opposed to continuous-time, there is a maximum frequency in discrete-time equal to π.

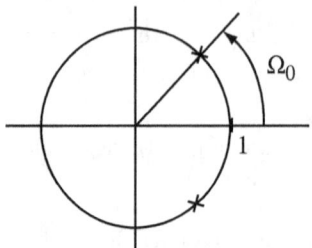

Figure 6.7: Poles of a sinusoidal signal

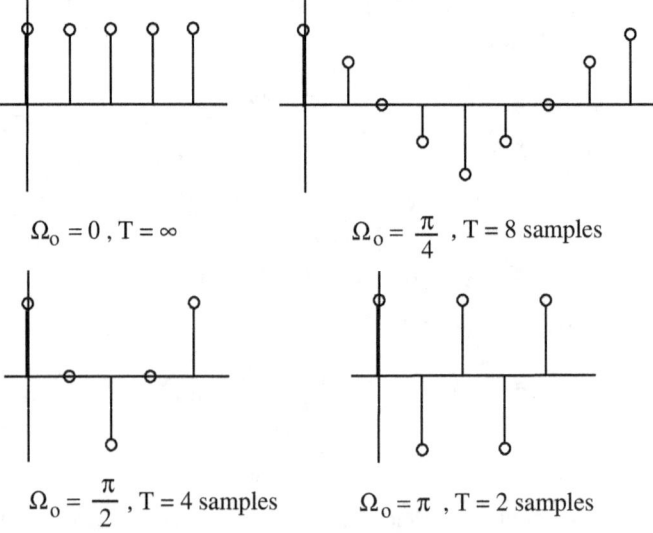

Figure 6.8: Discrete-time sinusoidal signals of different frequencies

6.1.3   The z-plane

Pole locations and signal shapes

As in continuous-time, the examples of simple z-transforms provide useful information about interpretations of the z-plane. The interpretations are different than in continuous-time, however. For signals with a single real pole $(X(z) = z/(z - p))$, or two complex poles $(X(z) = cz/(z - p) + c^*z/(z - p^*))$, the associated signals are shown in Fig. 6.9.

There are some strong similarities between the z−plane and the s−plane,

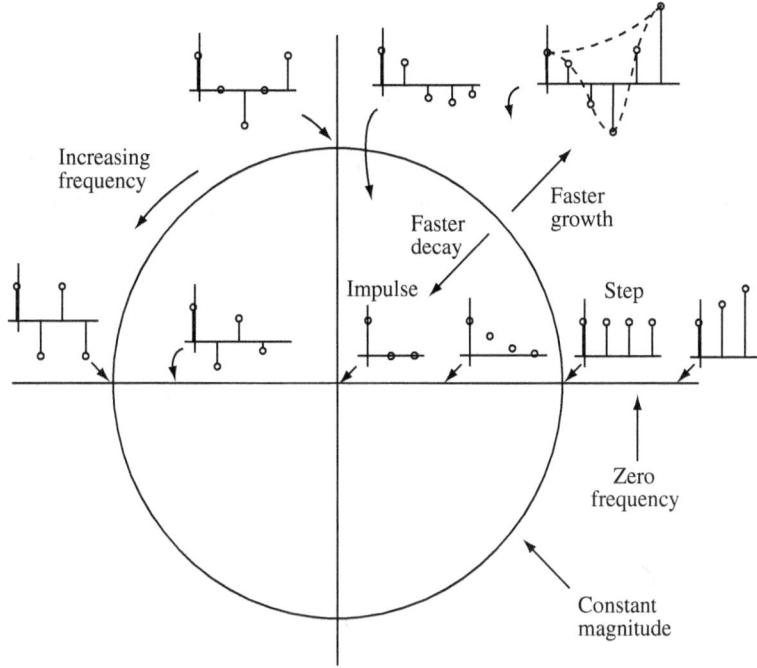

Figure 6.9: Signal characteristics as a function of pole location

with

$$|z| = 1 \text{ (unit circle)} \Leftrightarrow s = 0 \text{ } (j\omega - \text{axis})$$
$$|z| > 1 \text{ (outside the unit circle)} \Leftrightarrow \text{Re}(s) > 0 \text{ (open right half-plane)}$$
$$|z| < 1 \text{ (inside the unit circle)} \Leftrightarrow \text{Re}(s) < 0 \text{ (open left half-plane)}$$

As in continuous-time, one may define useful quantities

For $|p| < 1$, the time constant: $\tau_c = -\dfrac{1}{\ln|p|}$

For $|p| < 1$, the 2% decay time: $\tau_{2\%} = -\dfrac{4}{\ln|p|}$

For $|p| > 1$, the time to double: $\tau_{double} = \dfrac{2}{\ln|p|}$

For $|p| \approx 1$, a useful approximation is

$$\tau_c \simeq \frac{1}{1 - |p|}, \tag{6.25}$$

so that

$$|p| = 0.9 \Rightarrow \tau_c = 9.5 \approx 10 \text{ samples}$$
$$|p| = 0.99 \Rightarrow \tau_c = 99.5 \approx 100 \text{ samples}$$

The time constant should be multiplied by 4 to obtain the time needed for a decay of the signal to 2% of its original value. Specifically, $|p| = 0.99 \Rightarrow 4\tau_c = 400$, which means that

$$0.99^{400} = 0.018 \approx 2\% \tag{6.26}$$

In other words, an exponentially decaying signal whose transform has a single pole at $p = 0.99$ will decay to 2% of its original value within 400 discrete-time instants, or samples.

For complex poles, the period of oscillations is given by

$$T_{osc} = \frac{2\pi}{\angle p}. \tag{6.27}$$

The result is not necessarily an integer number of samples.

Lines of constant damping

In continuous-time, one defines the damping factor ζ of a pole p as the cos of the angle of the pole with respect to the negative real axis, or

$$\zeta = \frac{-\operatorname{Re}(p)}{|p|} = \frac{-\operatorname{Re}(p)}{\sqrt{\operatorname{Re}(p)^2 + \operatorname{Im}(p)^2}} \qquad \text{(continuous-time)}. \tag{6.28}$$

The dashed line of Fig. 6.10 is a line of constant damping in continuous-time. It is known that an angle greater than $45°$, or a damping factor smaller than 0.707, yields overshoot in the step response and peaking in the frequency response.

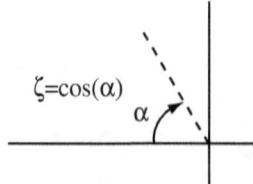

Figure 6.10: Definition of the damping factor ζ in continuous-time

Since the time constant and the period of oscillation associated with a complex pole are

$$\tau_c = -1/\operatorname{Re}(p), \qquad T_{osc} = 2\pi/\operatorname{Im}(p) \qquad \text{(continuous-time)}, \tag{6.29}$$

an interpretation of the damping factor is

$$\zeta = \frac{1/\tau_c}{\sqrt{(1/\tau_c)^2 + (2\pi/T_{osc})^2}} = \frac{1}{\sqrt{1 + (2\pi\tau_c/T_{osc})^2}}. \tag{6.30}$$

A desirable damping factor of $\zeta > 0.707$ means that $\tau_c < (T_{osc})/2\pi$, implying that the convergence time is small compared to the period of oscillation.

To obtain an equivalent result in discrete-time, we use the applicable definitions of time constant and of period of oscillations

$$\tau_c = -\frac{1}{\ln|p|}, \qquad T_{osc} = \frac{2\pi}{\angle p} \qquad \text{(discrete-time)}, \tag{6.31}$$

to obtain

$$\zeta = \frac{-\ln|p|}{\sqrt{(\ln|p|)^2 + (\angle p)^2}} \qquad \text{(discrete-time)}. \tag{6.32}$$

This equation can also be written as

$$\ln|p| = -\frac{\zeta}{\sqrt{1 - \zeta^2}} \angle p, \tag{6.33}$$

or

$$|p| = \frac{1}{e^{\alpha \angle p}} \quad \text{where } \alpha = \frac{\zeta}{\sqrt{1 - \zeta^2}}. \tag{6.34}$$

A curve of constant damping in discrete-time is the set of complex numbers p such that (6.34) is satisfied with a fixed ζ. (6.34) shows that the magnitude of the pole must decrease as the angle increases. A line of constant damping is a curve called a *logarithmic spiral*, and is shown in Fig. 6.11.

The line corresponding to $\zeta = 0.707$ is the curve

$$|p| = e^{-\angle p} \tag{6.35}$$

with $\angle p$ in radians. The curve defines the boundary separating well-damped from poorly-damped responses, in a similar way as the 45° line defines the same characteristics in the s-plane. In continuous-time, the line is such that the time constant is equal to the period of oscillation divided by 2π and the same remains true here.

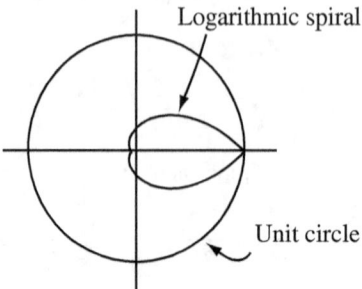

Figure 6.11: Line of constant damping for $\varsigma = 0.707$ (logarithmic spiral)

As an example of a signal with low damping, consider

$$p = 0.5 \pm j0.5 \Rightarrow |p| = 0.707, \ \angle p = \pi/4. \tag{6.36}$$

The damping factor can be computed from (6.32) to be $\zeta = 0.404$, which is low. Indeed, the time signal $p^k + (p^*)^k$ is shown in Fig. 6.12 and exhibits a visible oscillation.

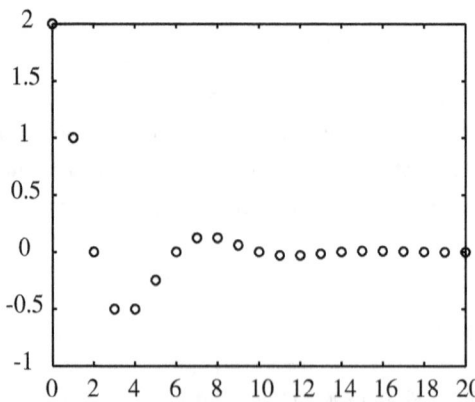

Figure 6.12: Discrete-time signal with low damping

Connections to the discrete-time Fourier transform

The *discrete-time Fourier transform* (DTFT) is defined to be

$$\bar{X}(\Omega) = \sum_{k=-\infty}^{\infty} x(k)e^{-j\Omega k}. \tag{6.37}$$

We use the notation $\bar{X}$ to distinguish the DTFT from the z-transform $X(z)$. The DTFT is defined on a finite interval (from $-\pi$ to π), or as a periodic function on an infinite interval. An example of a DTFT is shown in Fig. 6.13 (the magnitude is shown only, but the DTFT is generally a complex function).

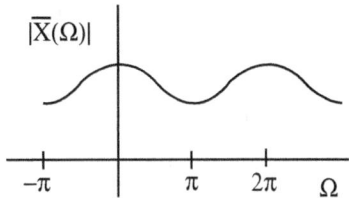

Figure 6.13: Discrete-time Fourier transform

One may observe that

$$\bar{X}(\Omega) = [X(z)]_{z=e^{j\Omega}} \qquad \text{if } x(k) = 0 \quad \text{for } k < 0. \tag{6.38}$$

In other words, we must assume that $x(k)$ is zero for negative time so that the bilateral and unilateral transforms can be related. The DTFT is equal to the z-transform *evaluated on the unit circle* ($z = e^{j\Omega}$). The property requires that the transforms exist in an ordinary sense on the unit circle, *i.e.*, that the region of convergence of the z-transform includes the unit circle. The property is not satisfied by a sinusoid, for which the region of convergence is $|z| > 1$

6.2   Properties of the z-transform

Properties can be established for the z-transform that are similar to those of the Laplace transform. The proofs are usually straightforward.

1. **Linearity**

$$x(k) = ax_1(k) + bx_2(k) \Leftrightarrow X(z) = aX_1(z) + bX_2(z). \tag{6.39}$$

2. **Right shift**

The *right shift* or *one-step delay* is defined by

$$y(k) = x(k-1)u(k-1), \qquad (6.40)$$

where $u(k)$ is the unit step. The signal and the one-step delayed signal are shown in Fig. 6.14. Since

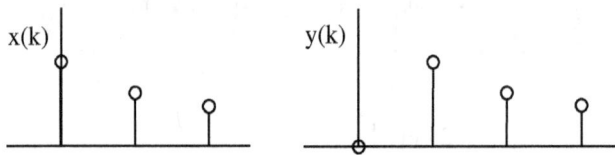

Figure 6.14: Right shift of a signal

$$X(z) = x(0) + x(1)z^{-1}\cdots, \qquad (6.41)$$

the output satisfies

$$Y(z) = x(0)z^{-1} + x(1)z^{-2} + \cdots = z^{-1}X(z). \qquad (6.42)$$

Therefore, we can associate

$$z^{-1} \Leftrightarrow \text{one step delay}, \qquad (6.43)$$

meaning that multiplication of the transform by z^{-1} is equivalent to a one-step delay of the signal. (6.40) assumed that $x(k)$ was multiplied by a unit step. Otherwise,

$$Y(z) = x(-1) + z^{-1}X(z). \qquad (6.44)$$

Examples of delayed signals are shown in Fig. 6.15 and include a delayed impulse and a delayed exponential signal. An observation is that a signal with a pole at $z = a$ and no zero at $z = 0$ is an exponential signal, but with a zero value at the initial time instant.

3. **Left shift**

In a similar manner, one can obtain the formula for a left shift

$$y(k) = x(k+1) \Rightarrow Y(z) = z\left(X(z) - x(0)\right). \qquad (6.45)$$

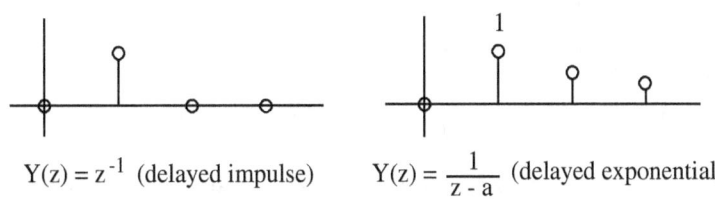

Figure 6.15: Delayed impulse and delayed exponential

To prove this fact, note that

$$
\begin{aligned}
Y(z) &= y(0) + y(1)z^{-1} + y(2)z^{-2} + \dots \\
&= x(1) + x(2)z^{-1} + x(3)z^{-2} + \dots \\
&= z\left(x(0) + x(1)z^{-1} + x(2)z^{-2} + x(3)z^{-3} + \dots\right) - zx(0).
\end{aligned}
$$

$$(6.46)$$

4. Initial value

The initial value of the signal $x(k)$ can be obtained as

$$x(0) = \lim_{z \to \infty} X(z). \tag{6.47}$$

The result follows from the fact that

$$\lim_{z \to \infty} X(z) = \lim_{z^{-1} \to 0} X(z) = \lim_{z^{-1} \to 0} \left(\sum_{k=0}^{\infty} x(k)z^{-k}\right) = x(0).$$

$$(6.48)$$

Note that by virtue of the definition of the z-transform, any transform $X(z)$ that is rational must be proper (degree numerator $\leqslant$ degree denominator). Therefore, the limit always exists. $x(0)$ is the ratio of the coefficients associated with the highest power of z in the numerator and in the denominator. For example

$$
\begin{aligned}
X(z) &= \frac{2z^2 + 1}{3z^2 + z} \Rightarrow x(0) = \frac{2}{3} \\
X(z) &= \frac{2z + 1}{3z^2 + z} \Rightarrow x(0) = 0.
\end{aligned}
$$

$$(6.49)$$

Other values of $x(k)$ may also be obtained in a similar manner, e.g.

$$x(1) = \lim_{z \to \infty} \left(zX(z) - x(0)\right). \tag{6.50}$$

5. **Final value**

If $\lim_{k \to \infty} x(k)$ exists, then

$$\lim_{k \to \infty} x(k) = \lim_{z \to 1}(z-1)X(z). \tag{6.51}$$

The result is similar to the equivalent result for the Laplace transform, but $s = 0$ is replaced by $z = 1$.

6. **Multiplication by time**

$$y(k) = kx(k) \Leftrightarrow Y(z) = -z\frac{d}{dz}X(z). \tag{6.52}$$

For example, consider the transform

$$x(k) = a^k \Leftrightarrow X(z) = \frac{z}{z-a}. \tag{6.53}$$

One may deduce the transforms of the signals

$$
\begin{aligned}
y_1(k) &= ka^k \Leftrightarrow Y_1(z) = -z\frac{(z-a)-z}{(z-a)^2} = \frac{az}{(z-a)^2}, \\
y_2(k) &= k^2 a^k \Leftrightarrow Y_2(z) = -z\frac{a(z-a)^2 - az\,2(z-a)}{(z-a)^4} = \frac{az(z+a)}{(z-a)^3}, \\
y_3(k) &= k^3 a^k \Leftrightarrow Y_3(z) = \cdots = \frac{az\,(z^2 + 4az + a^2)}{(z-a)^4}.
\end{aligned}
\tag{6.54}
$$

As in continuous-time, powers of time k^n are associated with poles of multiplicity $n+1$. However, numerator polynomials appear in the transforms that did not appear in the transforms in continuous-time. The time-domain signals associated to transforms with different polynomials in the numerators are given by more complicated functions of time. For example

$$Y(z) = \frac{z}{(z-a)^n} \Leftrightarrow y(k) = \frac{\overbrace{k(k-1)\,\cdots\,(k-n+2)}^{n-1 \text{ terms}}}{(n-1)!}a^{k-n+1}. \tag{6.55}$$

In particular

$$
\begin{aligned}
Y(z) &= \frac{z}{(z-a)^3} \Leftrightarrow y(k) = \frac{1}{2}k(k-1)a^{k-2}, \\
Y(z) &= \frac{z}{(z-a)^4} \Leftrightarrow y(k) = \frac{1}{6}k(k-1)(k-2)a^{k-3}. \tag{6.56}
\end{aligned}
$$

7. **Convolution**:

As in continuous-time, if

$$x(k) = 0 \text{ and } h(k) = 0 \quad \text{for} \quad k < 0, \tag{6.57}$$

then

$$y(k) = h(k) * x(k) \Leftrightarrow Y(z) = H(z)\, X(z). \tag{6.58}$$

In discrete-time, the convolution operation is given by

$$h(k) * x(k) = \sum_{i=-\infty}^{\infty} h(i)x(k-i). \tag{6.59}$$

Under the assumption of zero signals for negative time,

$$y(k) = \sum_{i=0}^{k} h(i)x(k-i). \tag{6.60}$$

Consider, for example, the convolution of a signal $x(k)$ with a step signal $u(k)$, yielding

$$y(k) = x(k) * u(k) = \sum_{i=0}^{k} x(i) \Leftrightarrow Y(z) = \frac{z}{z-1}X(z). \tag{6.61}$$

In other words,

$$\text{Discrete-time integrator} \Leftrightarrow \frac{z}{z-1}, \tag{6.62}$$

compared to $1/s$ in continuous-time. The formula for the integration of a signal could also have been derived quickly by transforming the recursion formula for $y(k)$, so that

$$y(k) = y(k-1) + x(k) \Rightarrow Y(z) = z^{-1}Y(z) + X(z)$$
$$\Rightarrow Y(z) = \frac{z}{z-1}X(z) \tag{6.63}$$

(assuming that $y(-1) = 0$). Note that, for a slightly different formulation

$$y(k) = y(k-1) + x(k-1) \Rightarrow Y(z) = z^{-1}Y(z) + z^{-1}X(z)$$
$$\Rightarrow Y(z) = \frac{1}{z-1}X(z) \tag{6.64}$$

(assuming that $y(-1) = x(-1) = 0$). In other words, multiplication by $1/(z-1)$, instead of $z/(z-1)$ also corresponds to an integrator, but with $x(k)$ delayed by one step.

6.3   Inversion of z-transforms

6.3.1   Inversion using partial fraction expansions

The method is similar to the one used for Laplace transforms. However, it turns out that it is more convenient to invert $X(z)/z$, rather than $X(z)$. Assume first that $X(z)$ has no repeated poles and no pole at $z = 0$. A partial fraction expansion gives

$$\frac{X(z)}{z} = \frac{c_0}{z} + \sum_{i=1}^{n} \frac{c_i}{z - p_i}, \tag{6.65}$$

with

$$
\begin{aligned}
c_0 &= X(0), \\
c_i &= \left[(z - p_i) \frac{X(z)}{z} \right]_{z=p_i} .
\end{aligned}
\tag{6.66}
$$

Then, one can write $X(z)$ as

$$X(z) = c_0 + \sum_{i=1}^{n} \frac{c_i z}{(z - p_i)}, \tag{6.67}$$

and obtain directly the time function

$$x(k) = c_0 \delta(k) + \sum_{i=1}^{n} c_i (p_i)^k. \tag{6.68}$$

Note that the preliminary division of $X(z)$ by z ensures that the function in the time domain can be obtained directly, without shifting. For complex poles, the two complex conjugate time-domain signals can be merged to produce a real signal

$$\frac{cz}{z - p} + \frac{c^* z}{z - p^*} \Leftrightarrow cp^k + c^* p^{*k} = 2|c| \, |p|^k \cos(\angle p \, k + \angle c). \tag{6.69}$$

Example: let

$$X(z) = \frac{1}{(z - 1)(z + 1)}. \tag{6.70}$$

Performing the partial fraction expansion

$$\frac{X(z)}{z} = \frac{1}{z(z - 1)(z + 1)} = -\frac{1}{z} + \frac{1/2}{z - 1} + \frac{1/2}{z + 1}, \tag{6.71}$$

so that

$$X(z) = -1 + \frac{1}{2} \frac{z}{z-1} + \frac{1}{2} \frac{z}{z+1}, \tag{6.72}$$

and the time-domain signal is obtained

$$x(k) = -\delta(k) + \frac{1}{2} + \frac{1}{2}(-1)^k. \tag{6.73}$$

The values of $x(k)$ are

$$x(0) = 0, \ x(1) = 0, \ x(2) = 1, \ x(3) = 0, \ x(4) = 1, \ \ldots \tag{6.74}$$

If $X(z)$ has poles at $z = 0$, or has other repeated poles, one must use the more general formula of partial fraction expansion. Poles at $z = 0$ will lead to terms of the form

$$\frac{c}{z^n} \Leftrightarrow c\delta(k - n), \tag{6.75}$$

i.e., to delayed impulses. Repeated poles other than at $z = 0$ generally are inverted using the formula

$$\frac{cz}{(z-p)^n} + \frac{c^*z}{(z-p^*)^n}$$
$$\Leftrightarrow 2|c| \frac{k(k-1) \ \cdots \ (k-n+2)}{(n-1)!} |p|^{k-n+1} \cos\left(\angle p \ (k-n+1) + \angle c\right). \tag{6.76}$$

While this is a more complicated expression than in continuous-time, it is no more difficult to apply and yields qualitatively similar results.

6.3.2 Inversion using long division

The z-transform permits an inversion procedure that does not have a useful equivalent in continuous-time. Essentially, it consists in reconstructing the power series $X(z) = x(0) + x(1)z^{-1} + x(2)z^{-2} \cdots$ by polynomial division (long division). For example, one can divide z by $z - a$ to reconstruct the signal associated with the transform

$$X(z) = \frac{z}{z-a}, \tag{6.77}$$

as shown in the following computation.

$$
\begin{array}{r}
1 + az^{-1} + a^2 z^{-2} \cdots \\
\hline
z - a \,\big|\; z \\
\underline{z - a} \\
a \\
\underline{a - a^2 z^{-1}} \\
a^2 z^{-1} \\
\underline{a^2 z^{-1} - a^3 z^{-2}} \\
a^3 z^{-2} \cdots
\end{array}
$$

$$\Rightarrow X(z) = 1 + az^{-1} + a^2 z^{-2} \cdots$$

$$\uparrow \quad \uparrow \quad \uparrow$$

$$x(0) \quad x(1) \quad x(2)$$

First example using long division

Another example is

$$X(z) = \frac{1}{z^2 - 1}, \tag{6.78}$$

which was inverted by partial fraction expansion in (6.74). The same result may be obtained through the following computation.

$$
\begin{array}{r}
z^{-2} + z^{-4} + z^{-6} + \cdots \\
\hline
z^2 - 1 \,\big|\; 1 \\
\underline{1 - z^{-2}} \\
z^{-2} \\
\underline{z^{-2} - z^{-4}} \\
z^{-4} \cdots
\end{array}
$$

$$\Rightarrow X(z) = z^{-2} + z^{-4} + z^{-6} + \cdots$$

$$\uparrow \quad \uparrow \quad \uparrow$$

$$x(2) \quad x(4) \quad x(6)$$

Second example using long division

6.3.3 Conclusions drawn from the procedure of partial fraction expansion

Conclusions can be derived from the known results of partial fraction expansions, as in continuous-time. Assuming a rational transform $X(z) = N(z)/D(z)$, with poles at $z = p_i$, one may immediately state that $x(k)$ is a linear combination of the following functions:

(a) $\delta(k)$ (an impulse).

(b) $\delta(k - l)$, $l = 1 \cdots n$ (delayed impulses), if $X(z)$ has n poles at $z = 0$.

(c) p_i^k if p_i is real.

(d) $k(k-1) \cdots (k - l + 2)p_i^k$, for $l = 2 \cdots n$,

   if p_i is real and is repeated n times.

(e) $|p_i|^k \cos(\angle p_i \, k + \phi_i)$ if p_i is complex for some ϕ_i.

(f) $k(k-1) \cdots (k-l+2)|p_i|^k \cos\left(\angle p_i \, k + \phi_{i,\,l}\right)$, for $l = 2 \cdots n$,

and for some $\phi_{i,l}$, if p_i is complex and is repeated n times.

Not all terms must be present, but one knows that the highest order terms (those corresponding to $l = n$) cannot have zero coefficients.

Example: consider a signal with double poles at $z = \pm j$, as shown in Fig. 6.16.

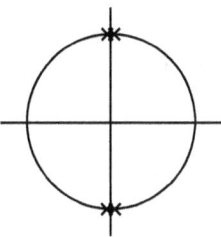

Figure 6.16: Example with double poles at $z = \pm j$

The signal $x(k)$ must be a linear combination of

$$\delta(k), \ \cos\left(\frac{\pi}{2}k + \phi_{11}\right), \ k\cos\left(\frac{\pi}{2}k + \phi_{12}\right), \tag{6.79}$$

and the coefficient multiplying the last term must be nonzero. The coefficient multiplying the first term is nonzero unless $X(z)$ is strictly proper.

6.3.4   Properties of signals

Properties of signals with rational transforms can be obtained as in continuous-time, with the equivalence

$$\begin{aligned} |z| &= 1 \Leftrightarrow \operatorname{Re}(s) = 0 \\ |z| &> 1 \Leftrightarrow \operatorname{Re}(s) > 0 \\ |z| &< 1 \Leftrightarrow \operatorname{Re}(s) < 0 \end{aligned} \tag{6.80}$$

As a consequence, the following properties of signals with rational transforms can be established

- $x(k)$ converges to zero $\Leftrightarrow$ all poles are inside the unit circle.

- $x(k)$ converges $\Leftrightarrow$ all poles are inside the unit circle, except at most a single pole at $z = 1$.

- $x(k)$ is bounded $\Leftrightarrow$ all poles are inside the unit circle or are non-repeated poles on the unit circle.

6.4   Discrete-time systems

6.4.1   Definition and examples

Discrete-time systems transform a discrete-time signal into another discrete-time signal. Linear time-invariant systems are characterized by their impulse response $h(k)$, whose z-transform $H(z)$ is the transfer function of the system. The most common discrete-time systems are described by *difference equations*, which have rational transfer functions.

Example 1: consider a bank deposit with daily interest such that

$$
\begin{aligned}
y(k) &= \text{balance at the end of day } k \\
x(k) &= \text{deposit on day } k \\
\alpha &= \text{daily interest rate}
\end{aligned}
\tag{6.81}
$$

Then, $y(k)$ satisfies a difference equation

$$y(k) = y(k-1) + \alpha y(k-1) + x(k), \tag{6.82}$$

so that the z-transform is given by

$$Y(z) = (1+\alpha)z^{-1}Y(z) + X(z). \tag{6.83}$$

Then

$$Y(z) = \frac{z}{z - (1+\alpha)}X(z), \tag{6.84}$$

so that the transfer function of the system is

$$H(z) = \frac{z}{z - (1+\alpha)}. \tag{6.85}$$

$H(z)$ is the z-transform of the impulse response

$$h(k) = (1+\alpha)^k. \tag{6.86}$$

This impulse response is shown in Fig. 6.17 and is unbounded. The money grows exponentially with a single initial deposit: the system is unstable!

Note that the daily rate can be transformed into a yearly rate (and vice-versa) with

$$3\% \text{ yearly rate} \Rightarrow (1+\alpha)^{365} = 1.03, \text{ or } \alpha = 8.1 \ 10^{-5}. \tag{6.87}$$

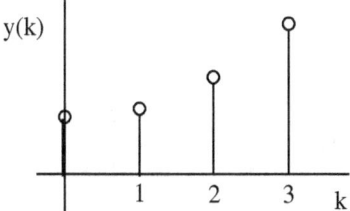

Figure 6.17: Impulse response of a first-order unstable system

The time to double can also be determined (assuming a 3% yearly rate)

$$(1 + \alpha)^k = 2 \Rightarrow k = \frac{\ln(2)}{\ln(1 + \alpha)} = 8.5592 \ 10^3 \text{ days} = 23.45 \text{ years.}$$
(6.88)

Example 2: consider the case of a microphone that receives the signal produced by a speaker. The system is shown in Fig. 6.18. There is a direct path from the speaker to the microphone, but a reflection is also received from a surface in the room. More echos may be present in the received signal.

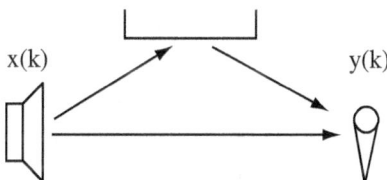

Figure 6.18: Transmission of sound with an echo

Fig. 6.19 shows the impulse response estimated experimentally in a laboratory system. Samples were taken at a rate of 10 kHz. The first part of the response exhibits a pair of large positive and negative pulses that may be attributed to the compression of air. A delay of 20 samples is visible and is associated with the distance from the speaker to the microphone. The 2 ms delay at a speed of sound of 1125 ft/s corresponds to a distance of 2.25 ft. Afterwards, the response decays to zero with some oscillations. In active noise control, it is common to treat the impulse response as equal to zero after some time. For example, the 300 samples shown on the figure may be taken to be the length of the impulse response of the system.

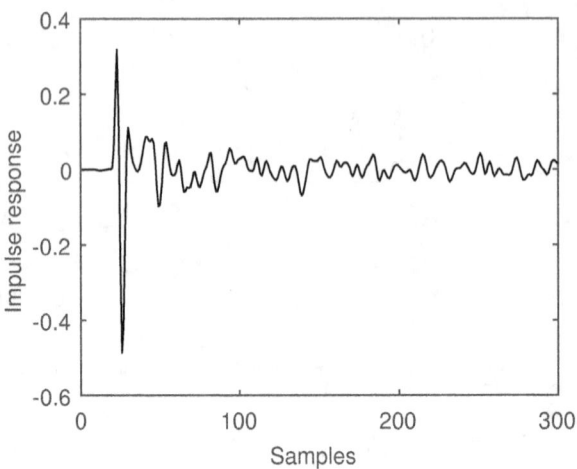

Figure 6.19: Impulse response corresponding to sound transmission in a room

6.4.2   FIR and IIR systems

FIR System (Finite Impulse Response)

A *finite impulse response* (FIR) system is such that the impulse response is a finite-time signal, *i.e.*, for some n

$$h(k) = 0 \qquad \text{for } k > n. \tag{6.89}$$

The transfer function of an FIR system is of the form

$$\begin{aligned} H(z) &= h(0) + h(1)z^{-1} + \cdots h(n)z^{-n} \\ &= \frac{h(0)z^n + \cdots + h(n)}{z^n}. \end{aligned} \tag{6.90}$$

In other words

$$\text{A system is FIR } \Leftrightarrow H(z) \text{ is rational with all poles at } z = 0 \tag{6.91}$$

Example 2 (echo system) was an example of an FIR system.

For an FIR system

$$\begin{aligned} y(k) &= h(k) * x(k) = \sum_{i=-\infty}^{\infty} h(i)x(k - i) \\ &= h(0)x(k) + h(1)x(k - 1) + \cdots + h(n)x(k - n). \end{aligned} \tag{6.92}$$

In other words, the output signal is the linear combination of the delayed values of the input signal, and is particularly easy to compute.

IIR System (Infinite Impulse Response)

Systems that are not FIR are *infinite impulse response* systems (IIR) and are such that

$$\text{For all } n, \text{ there exists } k \geqslant n \text{ such that } h(k) \neq 0. \tag{6.93}$$

Example 1 (bank deposit) was an example of an IIR system.

6.4.3 BIBO stability

BIBO stability is defined as for continuous-time systems, and stability conditions on rational transfer functions $H(z)$ can also be obtained in a similar manner, leading to the following condition:

$$H(z) \text{ is BIBO stable} \Leftrightarrow \text{all poles of } H(z) \text{ are inside the unit circle}$$

Note that *an FIR system is always stable.*

Examples

(a)

$$H(z) = \frac{1}{z(z - 0.5)} \quad \text{stable.} \tag{6.94}$$

(b)

$$H(z) = \frac{1}{z - 2} \quad \text{unstable.} \tag{6.95}$$

(c)

$$H(z) = \frac{1}{(z - 0.8 + j0.8)(z - 0.8 - j0.8)} \quad \text{unstable,} \tag{6.96}$$

since

$$|p|^2 = 0.8^2 + 0.8^2 = 1.28 \Rightarrow |p| > 1. \tag{6.97}$$

(d)

$$H(z) = \frac{1}{z^3} \quad \text{stable.} \tag{6.98}$$

(e)

$$H(z) = \frac{1}{z - 1} \quad \text{unstable.} \tag{6.99}$$

For cases (c) and (d), the output will be unbounded for most input signals. For case (e), the output will be unbounded for signals containing a DC component or bias.

6.4.4   Responses to step inputs

A step signal of arbitrary magnitude x_m has a transform

$$x(t) = x_m \Leftrightarrow X(z) = \frac{z}{z-1} \, x_m. \tag{6.100}$$

Therefore, the step response of a system with transfer function $H(z)$ is given by

$$Y(z) = H(z) \, \frac{z}{z-1} \, x_m. \tag{6.101}$$

As in continuous-time, one can predict the results that would be obtained if one performed a partial fraction expansion of $Y(z)/z$ (note the division by z to facilitate the expansion). If $H(z) = N(z)/D(z)$ is BIBO stable, the step response is composed of the steady-state response and the transient response, with

$$Y(z) = \underbrace{[H(z)]_{z=1} \, \frac{z}{z-1} \, x_m}_{\text{steady-state response}} + \underbrace{\frac{N_1(z)}{D(z)}}_{\text{transient response}}, \tag{6.102}$$

where $N_1(z)$ is a polynomial depending on $N(z)$ and $D(z)$. In the steady-state

$$\lim_{k \to \infty} y(k) = \lim_{z \to 1} (z-1) Y(z) = H(1) \, x_m. \tag{6.103}$$

The DC gain of a discrete-time system is $H(1)$.

6.4.5   Responses to sinusoidal inputs

We found earlier that the transform of a cos signal was given by

$$x(k) = \cos\left(\Omega_0 k\right) \Leftrightarrow X(z) = \frac{1}{2} \frac{z}{z - e^{j\Omega_0}} + \frac{1}{2} \frac{z}{z - e^{-j\Omega_0}} = \frac{z^2 - \cos \Omega_0 z}{z^2 - 2 \cos\left(\Omega_0\right) z + 1}. \tag{6.104}$$

The transform has two poles on the unit circle, placed at an angle Ω_0 from the real axis. Ω_0 is the frequency of the signal, in radians, with an associated period of $2\pi/\Omega_0$ in samples.

Using partial fraction expansions, the response of a BIBO stable system with rational transfer function $H(z)$ to an input $x(k) = x_m \cos\left(\Omega_0 k\right)$ is.

$$\begin{aligned}
Y(z) &= \frac{N(z)}{D(z)} \frac{z^2 - \cos\left(\Omega_0\right) z}{z^2 - 2 \cos\left(\Omega_0\right) z + 1} \, x_m \\[2mm]
&= \underbrace{\frac{N_1(z)}{z^2 - 2 \cos\left(\Omega_0\right) z + 1} \, x_m}_{\text{steady-state response } Y_{ss}(z)} + \underbrace{\frac{N_2(z)}{D(z)}}_{\text{transient response } Y_{tr}(z)}
\end{aligned} \tag{6.105}$$

As in continuous-time, it turns out that $N_1(z)$ is such that

$$y_{ss}(k) = M x_m \cos{(\Omega_0 k + \phi)}, \qquad (6.106)$$

where

$$M e^{j\phi} = [H(z)]_{z=p} \qquad \text{and } p = e^{j\Omega_0}. \qquad (6.107)$$

In other words, M and ϕ are the magnitude and angle of the transfer function evaluated on the unit circle

$$M = \left|H\left(e^{j\Omega_0}\right)\right|, \quad \phi = \angle H\left(e^{j\Omega_0}\right). \qquad (6.108)$$

As in continuous-time, $H\left(e^{j\Omega_0}\right)$ is called the *frequency response* of the system. It is also the DTFT of the impulse response of the system.

Example

Consider the FIR system with impulse response

$$h(k) = \delta(k) - \delta(k - 4), \qquad (6.109)$$

so that $y(k) = x(k) - x(k - 4)$. The computation of the output and an implementation of the system are shown in Fig. 6.20, where D represents a one-step delay.

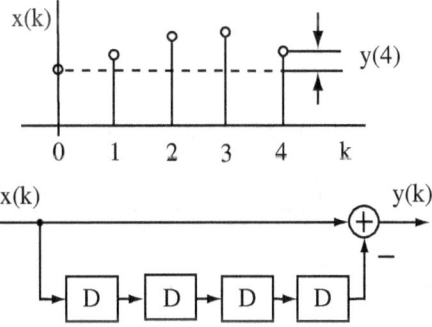

Figure 6.20: System with $H(z) = 1 - z^{-4}$, signal computation (top) and system implementation (bottom)

The system has a finite impulse response, and is therefore BIBO stable. Its transfer function is given by

$$H(z) = 1 - z^{-4} = \frac{z^4 - 1}{z^4}. \qquad (6.110)$$

There are 4 poles at $z = 0$, and 4 zeros at $z = 1$ (e^{j0}), $z = -1$ $(e^{j\pi})$, $z = j$ $(e^{j\pi/2})$, $z = -j$ $(e^{-j\pi/2})$. The four zeros on the unit circle imply that the frequency response is zero for the associated frequencies and, therefore, that the steady-state response is zero for the following signals:

$$x(k) = a \qquad\qquad \Omega_0 = 0$$
$$x(k) = a(-1)^k \qquad\qquad \Omega_0 = \pi$$
$$x(k) = a\cos\left(\tfrac{\pi}{2}k + \phi\right) \quad \Omega_0 = \pi/2$$

i.e., for signals with frequencies 0, $\pi/2$, and π. This result may be easily interpreted by considering the computation of the output signal in Fig. 6.20.

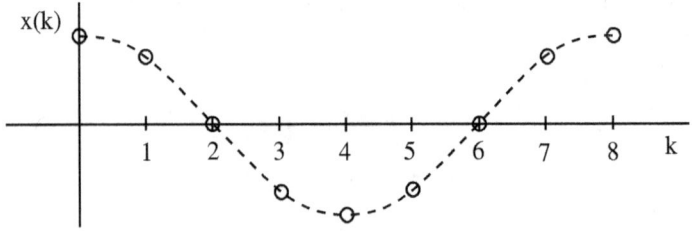

Figure 6.21: Sinusoidal signal with frequency $\Omega_0 = \pi/4$

On the other hand, consider the signal

$$x(k) = \cos\left(\frac{\pi}{4}k\right). \qquad\qquad (6.111)$$

The signal is shown in Fig. 6.21. The frequency response at $\Omega_0 = \pi/4$ is

$$H\left(e^{j\pi/4}\right) = \frac{\left(e^{j\pi/4}\right)^4 - 1}{\left(e^{j\pi/4}\right)^4} = \frac{e^{j\pi} - 1}{e^{j\pi}} = \frac{-2}{-1} = 2. \qquad (6.112)$$

Therefore, the steady-state output is given by

$$y_{ss}(k) = 2\cos\left(\frac{\pi}{4}k\right). \qquad\qquad (6.113)$$

As opposed to being eliminated, this signal is amplified by a factor of 2. Again, this result may be interpreted in view of the computation of the signal in Fig. 6.20.

As an additional example, consider

$$x(k) = \cos\left(\frac{\pi}{8}k\right), \qquad\qquad (6.114)$$

so that

$$H(e^{j\pi/8}) = \frac{j-1}{j} = 1 + j = \sqrt{2}\ e^{j45°} = \sqrt{2}\ e^{j\pi/4}, \tag{6.115}$$

and

$$y_{ss}(k) = \sqrt{2}\ \cos\left(\frac{\pi}{8}k + \frac{\pi}{4}\right). \tag{6.116}$$

6.4.6 Systems described by difference equations and effect of initial conditions

Difference equations are similar to input/output differential equations. As an example, consider the second-order equation

$$y(k) = -a_1 y(k-1) - a_0 y(k-2) + b_2 x(k) + b_1 x(k-1) + b_0 x(k-2). \tag{6.117}$$

At time $k = 0$, the recursion is started with initial conditions $y(-1)$, $y(-2)$, $x(-1)$, $x(-2)$, and the output signal at $k = 0$ is given by

$$y(0) = -a_1 y(-1) - a_0 y(-2) + b_2 x(0) + b_1 x(-1) + b_0 x(-2). \tag{6.118}$$

Typically, the initial conditions are set to zero, but the effect of nonzero values may be analyzed in much the same way as in continuous-time. For nonzero values prior to $k = 0$

$$y(k-1) \Leftrightarrow z^{-1}Y(z) + y(-1), \tag{6.119}$$

so that

$$y(k-2) \Leftrightarrow z^{-1}\left(z^{-1}Y(z) + y(-1)\right) + y(-2) = z^{-2}Y(z) + z^{-1}y(-1) + y(-2). \tag{6.120}$$

In this manner, the shifting formula can be extended to arbitrary time shifts, and used to account for arbitrary initial conditions.

Applying the formulas to the difference equation, one obtains

$$\begin{aligned}
Y(z) = \ & -a_1 z^{-1}Y(z) - a_1 y(-1) - a_0 z^{-2}Y(z) - a_0 z^{-1}y(-1) - a_0 y(-2) \\
& + b_2 X(z) + b_1 z^{-1}X(z) + b_1 x(-1) \\
& + b_0 z^{-2}X(z) + b_0 z^{-1}x(-1) + b_0 x(-2),
\end{aligned} \tag{6.121}$$

and

$$Y(z) \;=\; \underbrace{\frac{N(z)}{z^2 + a_1 z + a_0}}_{\substack{\text{Response to the initial conditions} \\ \text{or zero-input response } Y_{zi}(z)}}$$

$$+ \underbrace{\frac{b_2 z^2 + b_1 z + b_0}{z^2 + a_1 z + a_0} X(z)}_{\substack{\text{Response to the input} \\ \text{or zero-state response } Y_{zs}(z)}} , \qquad (6.122)$$

where

$$N(z) \;=\; -a_1 z^2 y(-1) - a_0 z y(-1) - a_0 z^2 y(-2)$$
$$+ b_1 z^2 x(-1) + b_0 z x(-1) + b_0 z^2 x(-2). \qquad (6.123)$$

The first term $Y_{zi}(s)$ is the response to the initial conditions (the zero-input response) and the second term $Y_{zs}(s)$ is the response to the input (the zero-state response). The two denominators are identical, and determine the poles of the system. The transfer function of the system is the rational function

$$H(z) = \frac{b_2 z^2 + b_1 z + b_0}{z^2 + a_1 z + a_0}. \qquad (6.124)$$

and can be directly transcribed from the original difference equation.

6.4.7   Internal stability definitions and properties

Internal stability definitions and properties follow as in continuous-time for a system described by a difference equation. The definitions are:

Asymptotically stable	$y_{zi}(k) \to 0$ as $k \to \infty$
Marginally stable	$y_{zi}(k)$ bounded
Internally unstable	$y_{zi}(k)$ unbounded

while the tests on the poles become

Asymptotically stable	All poles inside the unit circle
Marginally stable	All poles inside the unit circle + possible non-repeated poles on the unit circle
Internally unstable	At least one pole outside the unit circle or repeated pole on the unit circle

As in continuous-time, asymptotic stability is equivalent to bounded-input bounded-output stability. Note that an FIR system has all poles at $z = 0$, so that the response to initial conditions becomes exactly zero in finite time. An FIR system is always BIBO stable and asymptotically stable.

6.4.8   Realization of discrete-time transfer functions

FIR system

An FIR transfer function is of the form

$$H(z) = h(0) + h(1)z^{-1} + h(2)z^{-2} \cdots + h(n)z^{-n}, \qquad (6.125)$$

and the output may be computed using the formula

$$y(k) = h(0)x(k) + h(1)x(k-1) + \cdots + h(n)x(k-n). \qquad (6.126)$$

The implementation is shown in Fig. 6.22.

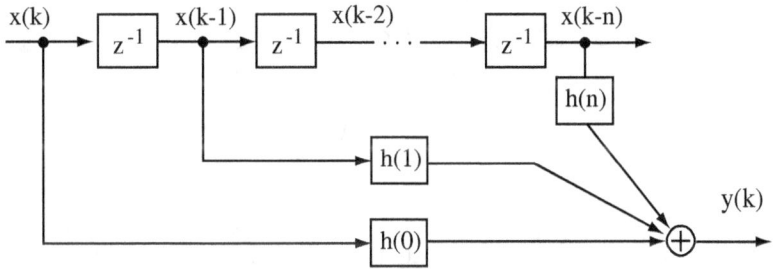

Figure 6.22: Implementation of an FIR transfer function

The operator z^{-1} is sometimes replaced by D, for delay. It is a one-step delay, and is realized with one memory location in a computer code. Specifically, code for an implementation of the FIR filter is of the form

```
Begin
y = h(0) * x + h(1) * m₁ + · · · + h(n) * mₙ
mₙ = mₙ₋₁
⋮
m₂ = m₁
m₁ = x
Repeat
```

where m_1, ..., m_n are n storage locations. The fact that the zero-input response vanishes in n steps means that the effect of any value placed in the registers m_1, ..., m_n, disappears n time instants after the system is started.

IIR system

An IIR system has the general transfer function

$$H(z) = \frac{b_n z^n + \cdots + b_0}{z^n + a_{n-1} z^{n-1} + \cdots + a_0}, \tag{6.127}$$

which may be translated into the difference equation

$$y(k) = -a_{n-1} y(k-1) - \ldots - a_0 y(k-n) + b_n x(k) + \ldots + b_0 x(k-n). \tag{6.128}$$

A direct implementation of the difference equation is shown in Fig. 6.23 and requires $2n$ storage elements.

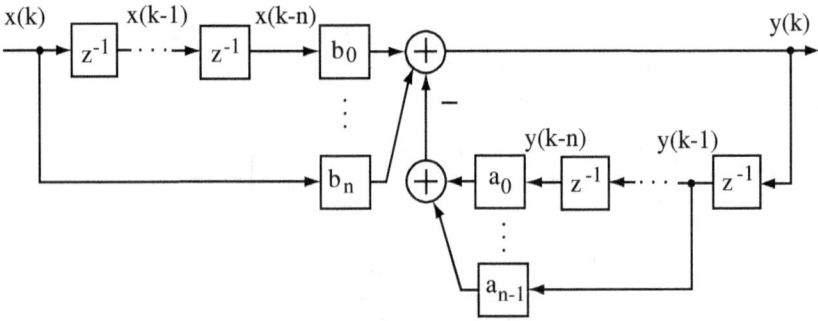

Figure 6.23: Direct implementation of an IIR transfer function

An alternate implementation is shown in Fig. 6.24. It only requires n storage elements, which is the minimum possible. The realization is equivalent to the canonical form discussed in continuous-time, and assumes that $b_n = 0$. To show that the realization implements the desired transfer function, note that

$$X_2(z) = z X_1(z), \ X_3(z) = z X_2(z), \ \ldots, \ X_n(z) = z X_{n-1}(z), \tag{6.129}$$

and

$$X(z) = z X_n(z) + a_0 X_1(z) + \cdots + a_{n-1} X_n(z). \tag{6.130}$$

Therefore

$$X_3(z) = z^2 X_1(z), \ \cdots, \ X_n(z) = z^{n-1} X_1(z), \tag{6.131}$$

and

$$X(z) = \left(z^n + a_{n-1}z^{n-1} + \cdots + a_0\right) X_1(z). \tag{6.132}$$

Also

$$
\begin{aligned}
Y(z) &= b_{n-1}X_n(z) + \cdots + b_0 X_1(z) \\
&= \left(b_{n-1}z^{n-1} + \cdots + b_0\right) X_1(z) \\
&= \frac{b_{n-1}z^{n-1} + \cdots + b_0}{z^n + a_{n-1}z^{n-1} \cdots + a_0} X(z).
\end{aligned}
\tag{6.133}
$$

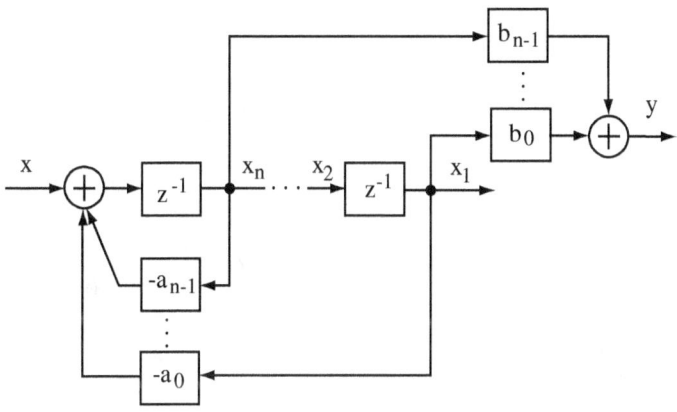

Figure 6.24: Minimal implementation of an IIR transfer function

As in continuous-time, if $b_n \neq 0$, the transfer function can be split into a gain and a strictly proper transfer function

$$H(z) = b_n + H_{sp}(z), \tag{6.134}$$

where $H_{sp}(z)$ is of the form

$$H_{sp}(z) = \frac{b'_{n-1}z^{n-1} + \cdots + b'_0}{z^n + a_{n-1}z^{n-1} + \cdots + a_0}. \tag{6.135}$$

Then, the output y is given by

$$y(k) = b_n x(k) + y'(k), \tag{6.136}$$

where $y'(k)$ is the output of a strictly proper system as shown in Fig. 6.24.

Realizability: to be *realizable*, a transfer function must be proper. With a unilateral z-transform (where the summation is from $k = 0$ to ∞), a system with a non-proper transfer function is not possible. With a bilateral z-transform (where the summation is from $k = -\infty$ to ∞), a non-proper transfer function is possible, but is still not realizable. In particular, the impulse response of $H(z) = z$ is an impulse occurring at $k = -1$. A system whose output responds before the input is applied is called non-*causal*, and does not respect cause-and-effect properties of physical systems (referred to as *causality*).

6.4.9   State-space models

A discrete-time state-space model has the form

$$\begin{aligned} x(k+1) &= Ax(k) + Bu(k) \\ y(k) &= Cx(k) + Du(k). \end{aligned} \tag{6.137}$$

The minimal implementation of the IIR system is an example of a state-space system with matrices

$$A = \begin{pmatrix} 0 & 1 & 0 & \cdots & 0 \\ 0 & 0 & 1 & \cdots & 0 \\ \vdots & & & & \\ -a_0 & & \cdots & & -a_{n-1} \end{pmatrix} \qquad B = \begin{pmatrix} 0 \\ \vdots \\ 0 \\ 1 \end{pmatrix}$$

$$C = \begin{pmatrix} b_0 & \cdots & b_{n-1} \end{pmatrix} \qquad\qquad D = 0. \tag{6.138}$$

As a result, a rational and proper transfer function can always be realized in discrete-time using a state-space description. Conversely, the transfer function may be computed by obtaining the output of the system using

$$\begin{aligned} zX(z) - x(0) &= AX(z) + BU(z) \\ Y(z) &= CX(z) + DU(z), \end{aligned} \tag{6.139}$$

which gives

$$Y(z) = \underbrace{\left(C(zI - A)^{-1}B + D\right)}_{H(z)} U(z) + \underbrace{C(zI - A)^{-1}x(0)}_{\text{Response to IC's}}. \tag{6.140}$$

$H(z)$ is the transfer function of the system and IC's means initial conditions. The poles are the eigenvalues of the matrix A, *i.e.*, the roots of $\det(zI - A)$.

In discrete-time, the general solution of the difference equation is relatively simple. Iterating on the recursion equation gives

$$
\begin{aligned}
x(1) &= Ax(0) + Bu(0) \\
x(2) &= Ax(1) + Bu(1) = A^2 x(0) + ABu(0) + Bu(1) \\
x(k) &= A^k x(0) + \sum_{i=0}^{k-1} CA^{k-1-i} Bu(i).
\end{aligned}
\tag{6.141}
$$

Therefore, the general solution is

$$
y(k) = \underbrace{CA^k x(0)}_{\substack{\text{Response to IC's} \\ \text{or zero-input response } Y_{zi}(z)}}
$$

$$
+ \underbrace{\sum_{i=0}^{k-1} \left(CA^{k-1-i} Bu(i) \right) + Du(k)}_{\substack{\text{Response to the input} \\ \text{or zero-state response } Y_{zs}(z)}}.
\tag{6.142}
$$

6.4.10   Extensions of other continuous-time results

The similarity between the z-transform and the Laplace transform implies that many results discussed earlier can be adapted to discrete-time. When differences arise, it is usually because of the distinctions between the stability regions for the continuous-time and discrete-time cases.

(a) Interconnected systems

The transfer functions of interconnected systems can be obtained exactly as in continuous-time. For example, the closed-loop system of Fig. 6.25 has transfer function $G(z)/(1 + G(z))$.

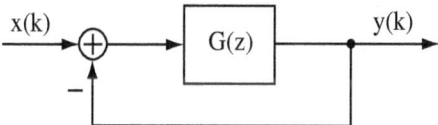

Figure 6.25: Discrete-time feedback system

(b) Desirable pole locations

The desirable region for the location of the closed-loop poles is the grey portion of the z-plane shown in Fig. 6.26. It is the intersection of the objectives of damping and settling time.

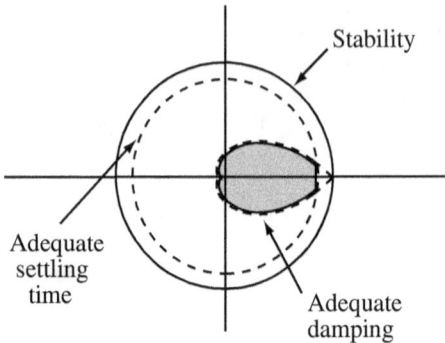

Figure 6.26: Desirable z-domain pole locations

(c) Routh-Hurwitz criterion

The Routh-Hurwitz criterion is not directly useful in discrete-time, because the objective is not to place the poles in the left half-plane anymore. The *Jury test* may be used to find the number of poles of a polynomial located outside the unit circle, and is the equivalent of the Routh-Hurwitz criterion.

(d) Root-locus

The poles of the feedback system of Fig. 6.25 can be obtained from the poles and zeros of the open-loop system using the *same* procedure as in continuous-time. Differences arise only with the interpretation of the results. In discrete-time, any system with a strictly proper transfer function becomes unstable in closed-loop for large gains. Even a stable first-order system with transfer function $b/(z - a)$ becomes unstable for large gain, which is not the case in continuous-time.

(e) Bode plots

Frequency response plots have the same significance in discrete-time as they have in continuous-time. The plots of the magnitude and phase of $H(e^{j\Omega})$ characterize the amplification and delay of the response to an input signal $\cos(\Omega t)$. It is only necessary to plot for a range $\Omega : 0 \to \pi$ on the x-axis. There is no simple method to draw these plots as for the Bode plots, and they are normally generated numerically.

(f) Nyquist criterion

The Nyquist criterion can be applied to test the stability of a closed-loop system, but a different curve must be used in discrete-time. In continuous-time, the open-loop transfer function is evaluated along the imaginary axis (completed at infinity), which is the portion of the s-plane that is associated with the sinusoidal response, as well as the boundary for stability (see Fig. 5.35). In discrete-time, the computation must be performed for

$$z = e^{j\Omega}, \quad \Omega = -\pi \rightarrow \pi, \tag{6.143}$$

as shown Fig. 6.27. Evaluating $H(z)$ along the Nyquist curve is equivalent to plotting the frequency response $H(e^{j\Omega})$ in the complex plane. Counting the number of encirclements of the $(-1, 0)$ point gives the number of unstable closed-loop poles, as in continuous-time.

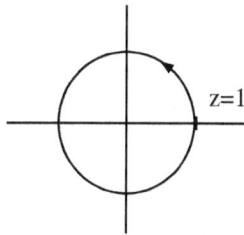

Figure 6.27: Nyquist curve in discrete-time

6.4.11   Example of root-locus in discrete-time

Consider a plant

$$P(z) = \frac{1}{z-1}, \tag{6.144}$$

which is an integrator with a unit step delay. The controller also has a pole at $z = 1$ and is given by

$$C(z) = g\frac{z - z_a}{z - 1}. \tag{6.145}$$

The closed-loop poles are the roots of

$$d_{CL}(z) = z^2 + (g-2)z + 1 - gz_a, \tag{6.146}$$

and the root-locus is shown on Fig. 6.28 for $z_a = 0.8$.

The root-locus is the same as it would be in continuous-time, but the conditions for stability are different. There is a value of the gain g_{max} for which one of the poles is equal to -1. The system is only stable up to $g = g_{max}$. For $g_0 = 1/z_a$, one of the poles is equal to zero. It is counterproductive to increase the gain beyond g_0.

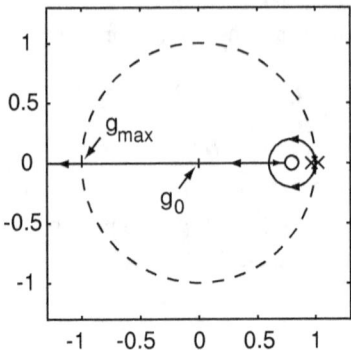

Figure 6.28: Root-locus for the discrete-time example (the unit circle is shown as a dashed curve)

A possible design consists in choosing a desired value for the closed-loop poles equal to some z_d, with $|z_d| < 1$. Then,

$$d_{CL}(z) = (z - z_d)^2 = z - 2z_d \, z + z_d^2 \tag{6.147}$$

is obtained for

$$g = 2(1 - z_d), \quad z_a = \frac{1 + z_d}{2}. \tag{6.148}$$

Such a design technique is referred to as *pole placement*. When both poles are placed at the same location, the choice corresponds to the breakaway point of the root-locus. The method can work well if reasonable values of z_d are chosen, which are generally values slightly smaller than 1.

6.5   Problems

Problem 6.1: (a) Find $x(0)$ if the z-transform of $x(k)$ is $X(z) = \frac{az - 1}{z - 1}$.
(b) Find $x(0)$ if the z-transform of $x(k)$ is $X(z) = \frac{z}{z^2 - az + a^2}$.

Problem 6.2: (a) Consider the *Newton-Raphson* method to find the zeros of a function $f(x)$, which is described by the difference equation

$$x(k) = x(k - 1) - \frac{f(x(k - 1))}{f'(x(k - 1))}, \tag{6.149}$$

where $f'(x(k-1))$ is the derivative of $f(x)$ evaluated at $x(k-1)$. Let $f(x) = ax^2$. Find the z-transform of $x(k)$ as a function of $x(-1)$, and give conditions under which $x(k)$ converges to zero as $k \to \infty$.

(b) Repeat part (a) for the *gradient* algorithm

$$x(k) = x(k-1) - f'(x(k-1)).\qquad(6.150)$$

Problem 6.3: (a) Use partial fraction expansions to find the $x(k)$ whose z-transform is $X(z) = \dfrac{1}{(z-1)(z-2)}$.

(b) Use partial fraction expansions to find the $x(k)$ whose z-transform is $X(z) = \dfrac{z}{z^2 - 2z + 2}$.

Problem 6.4: (a) Sketch the time function $x(k)$ that you would associate with the following poles: $p_1 = 0.9j$, $p_2 = -0.9j$. Only a sketch is required, but be as precise as possible.

(b) Repeat part (a) for: $p_1 = 1$, $p_2 = -1$.

(c) Repeat part (a) for: $p_1 = 0.3$, $p_2 = 0.9$.

(d) Repeat part (a) for: $p_1 = e^{j\pi/6}$, $p_2 = e^{-j\pi/6}$.

The pole locations for the 4 parts are shown in Fig. 6.29.

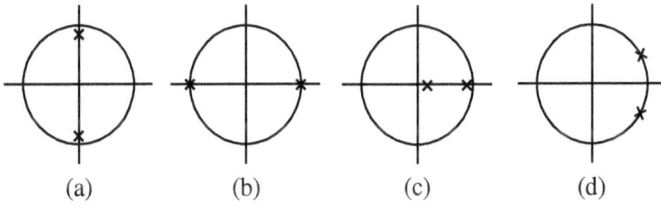

Figure 6.29: Pole locations for problem 6.4

Problem 6.5: (a) Find $Y(z)$ as a function of $X(z)$ if $y(k) = (-1)^k x(k)$. Assuming that $\bar{X}(\Omega)$, the DTFT of $x(k)$, is as shown on Fig. 6.30, sketch $\bar{Y}(\Omega)$, the DTFT of $y(k)$.

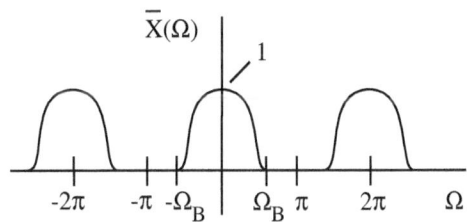

Figure 6.30: DTFT for problem 6.5

(b) Repeat part (a) for

$$y(k) = x(k/2) \quad \text{if n is even}$$
$$y(k) = 0 \qquad\quad \text{if n is odd}$$

Problem 6.6: (a) Using partial fraction expansions, find the signal $x(k)$ whose z-transform is

$$X(z) = \frac{4}{(z-1)(z^2+1)}. \tag{6.151}$$

Use the result to compute $x(k)$ for $k = 0, ..., 8$.

(b) Using long division, obtain $x(k)$ for the signal of part (a) and $k = 0, ..., 8$. Compare the results to those obtained in part (a).

Problem 6.7: For the signals whose z-transforms are given below, indicate whether the time functions $x(k)$ are bounded, converge to some value, or vanish in finite time.

(a) $X(z) = \dfrac{(z+1)}{(z+0.5)(z-0.7+0.7j)(z-0.7-0.7j)}$.

(b) $X(z) = (1 - 2z^{-1})(1 + 3z^{-1})$.

(c) $X(z) = \dfrac{(z-1)}{(z+1)(z+0.5)^2}$.

(d) $X(z) = \dfrac{(z+1)}{(z-1)(z+0.5)^2}$.

(e) $X(z) = \dfrac{(z+1)}{z(z-1)}$.

(f) $X(z) = \dfrac{z^{10}}{(z+5)}$.

(g) $X(z) = \dfrac{(z+1)^2}{(z^2+1)(z-0.5)}$.

(h) $X(z) = \dfrac{(z-2)^2}{z^3(z-1)}$.

Problem 6.8: Indicate whether the discrete-time systems with the following transfer functions are BIBO stable.

(a) $H(z) = \dfrac{z}{z-0.5}$.

(b) $H(z) = \dfrac{z^3}{(z^2+0.81)^2}$.

(c) $H(z) = \dfrac{z}{(z+1)(z+2)}$.

(d) $H(z) = \dfrac{z-10}{z^{10}}$.

(e) $H(z) = \dfrac{z+0.5}{(z+1)(z+0.25)}$.

(f) $H(z)$ corresponding to the difference equation: $y(k+1) - \frac{1}{2}y(k) = x(k+1) - 2x(k)$.

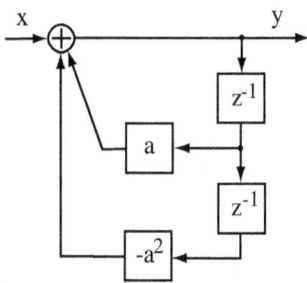

Figure 6.31: System for problem 6.9 (a)

Problem 6.9: (a) Find the transfer function $H(z) = Y(z)/X(z)$ and a condition on a such that the system of Fig. 6.31 is BIBO stable.
(b) Find the transfer function $H(z) = Y(z)/X(z)$ and indicate whether the system of Fig. 6.32 is BIBO stable.

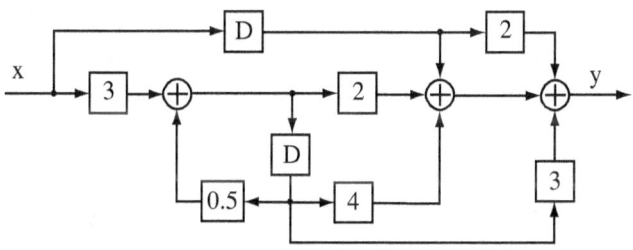

Figure 6.32: System for problem 6.9 (b)

Problem 6.10: (a) Find the transfer function $H(z) = Y(z)/X(z)$ corresponding to the difference equation

$$y(k) = y(k-1) + y(k-2) + x(k). \tag{6.152}$$

Is the system stable?
(b) Let the input be $x(k) = \delta(k)$, the discrete-time impulse. Show that the response of the system of part (a) with zero initial conditions is such that $\lim_{k \to \infty} \frac{y(k)}{y(k-1)}$ exists, and give its value.

Problem 6.11: (a) Find the steady-state response of the system with transfer function $H(z) = \frac{z-2}{z^4(z+0.5)}$ and an input $x(k) = 3$. Do not perform a partial fraction expansion: use the DC gain.

(b) Repeat part (a) for $x(k) = \cos(\pi k/2)$. Do not perform a partial fraction expansion: use the frequency response.

(c) Write a program (in Matlab, for example) to check the results of parts (a) and (b). Plot the input $x(k)$ and the output $y(k)$ for both cases over 40 time steps. To implement $H(z)$, find a difference equation that corresponds to the given transfer function, and let all initial conditions be zero.

Problem 6.12: Consider the system of Fig. 6.33, with $C(z) = g/(z-a)$ and $P(z) = 1/z$. Find conditions on g and a such that the steady-state error $e_{ss} = \lim_{k\to\infty} e(k)$ is zero for all constant inputs $r(k) = r_m$. Assume that $g > 0$.

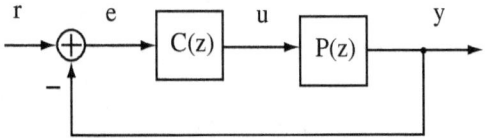

Figure 6.33: System for problem 6.12

Problem 6.13: (a) Find the initial value and the final value of the signal whose z-transform is

$$X(z) = \frac{z^2}{(z+0.5+0.9j)(z+0.5-0.9j)}. \tag{6.153}$$

(b) Find the initial value and the final value of the *step response* of the system

$$H(z) = \frac{2z^2 - 0.3z + 0.25}{z^2 - 0.6z + 0.25}. \tag{6.154}$$

Problem 6.14: (a) Sketch the time function $x(k)$ that you would associate with the following poles: $p_1 = 1$, $p_2 = j$, $p_3 = -j$. Only a sketch is required, but be as precise as possible.

(b) Repeat part (a) for: $p_1 = e^{j2\pi/3}$, $p_2 = e^{-j2\pi/3}$.

(c) Is a discrete-time system with the following poles BIBO stable?
$p_1 = 0.9 + 0.1j$, $p_2 = 0.9 - 0.1j$, $p_3 = -0.9 + 0.1j$, $p_4 = -0.9 - 0.1j$,
$p_5 = 0.1 + 0.9j$, $p_6 = 0.1 - 0.9j$, $p_7 = -0.1 + 0.9j$, $p_8 = -0.1 - 0.9j$.

(d) Repeat part (c) for: $p_1 = -1$, $p_2 = -0.5$, $p_3 = -0.5+0.5j$, $p_4 = -0.5-0.5j$.
Pole locations for the 4 parts are shown on Fig. 6.34.

Problem 6.15: (a) Consider the discrete-time system with transfer function

$$H(z) = \frac{z^4}{z^4 - 1}. \tag{6.155}$$

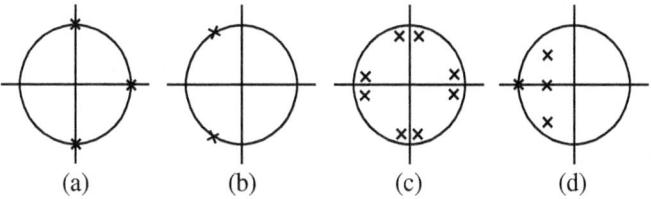

Figure 6.34: Pole locations for problem 6.14

What are the poles and zeros of the system? Is the system BIBO stable?
(b) Find the impulse response of the system of part (a) using long division.
Problem 6.16: (a) Using long division, find $y(0)$, $y(1)$, and $y(2)$ for

$$Y(z) = \frac{2z^3 + 13z^2 + z}{z^3 + 7z^2 + 2z + 1}. \qquad (6.156)$$

(b) Find the transfer function $H(z) = Y(z)/X(z)$ for the system of Fig. 6.35.
Give the values of the closed-loop poles and the range of gain g for which the
system is closed-loop stable. Show the root-locus of the system, with g being
the parameter that varies from 0 to ∞.

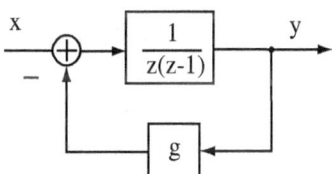

Figure 6.35: System for problem 6.16 (b)

Problem 6.17: (a) Let

$$y(n) = y(n-1) + \frac{3}{4}y(n-2) + x(n-1). \qquad (6.157)$$

Find the response $Y(z)$ to arbitrary initial conditions and an arbitrary input
$X(z)$. Determine whether the system is BIBO stable.
(b) Using partial fraction expansions, find the signal $y(k)$ for $y(-1) = 0$, $y(-2) =$
1, and an input $x(k)$ that is zero everywhere except $x(0) = 1$.

Problem 6.18: (a) Find the step response of the system

$$H(z) = \frac{1}{z\left(z^2 - z + 1/2\right)}. \qquad (6.158)$$

Indicate what the transient response and the steady-state response are. Compare the steady-state value to the value predicted by the DC gain.

(b) Find the steady-state response of the system of part (a) to $x(k) = \cos(\pi k/2)$ as well as for $x(k) = \cos(\pi k)$.

Problem 6.19: (a) Using partial fraction expansions, find the signal $x(k)$ whose z-transform is

$$X(z) = \frac{1}{(z-1)(z^2 - 2z + 2)}. \tag{6.159}$$

(b) Find the transfer function $H(z) = Y(z)/X(z)$ and a condition on a such that the system of Fig. 6.36 is BIBO stable.

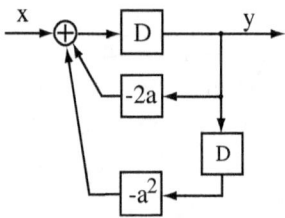

Figure 6.36: System for problem 6.19 (b)

Problem 6.20: Find the steady-state response of the system with transfer function $H(z) = \frac{z^4 - 1}{z^8}$ and an input $x(k) = \cos(\pi k/4)$. Do not perform a partial fraction expansion: use the frequency response. Plot the response $y_{ss}(k)$ (label the axes precisely).

Chapter 7

Sampled-data systems

7.1 Conversion continuous-time signal to discrete-time signal

7.1.1 Definition of sampling

Consider the transformation of a continuous-time signal $x(t)$ to a discrete-time signal $x_d(k)$, such that

$$x_d(k) = x(kT), \tag{7.1}$$

where T is called the *sampling period* (in seconds), $f_s = 1/T$ and $\omega_s = 2\pi/T$ are the *sampling frequency* (in Hz and in rad/s, respectively). The continuous/discrete conversion, is referred to as *sampling*, or *discretization*, and is shown in Fig. 7.1 for an arbitrary signal x(t). The operation is performed in analog-to-digital (A/D) converters. However, such converters also perform a *quantization* operation, which approximates real numbers by a finite set of numbers, coded by bits. The effects of quantization are ignored in the following discussion.

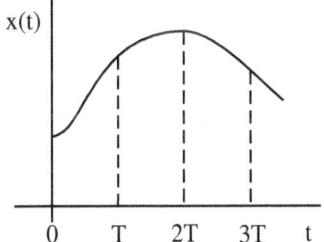
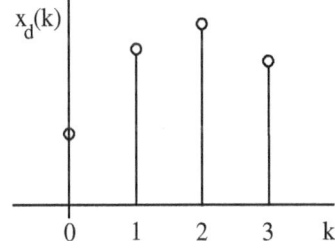

Figure 7.1: Conversion of a continuous-time signal to a discrete-time signal

Let $X(s)$ denote the Laplace transform of $x(t)$ and $X_d(z)$ denote the z-transform of $x_d(k)$. A natural question to ask is: how is $X_d(z)$ related to $X(s)$? The answer to the question turns out to be relatively simple when $X(s)$ is a rational function of s, but quite complicated in general.

7.1.2   Transform of a sampled signal with rational transform

Consider a general rational transform $X(s)$ with non-repeated poles

$$X(s) = \frac{N(s)}{D(s)} = \sum_{i=1}^{n} \frac{c_i}{s - p_i}, \tag{7.2}$$

where p_i are the poles of $X(s)$ and c_i are the coefficients of the partial fraction expansion. The corresponding signal is given by

$$x(t) = \sum_{i=1}^{n} c_i e^{p_i t}. \tag{7.3}$$

If we sample the signal every T seconds, the resulting discrete-time signal is given by

$$x_d(k) = \sum_{i=1}^{n} c_i e^{p_i k T}, \tag{7.4}$$

and the z-transform of the discrete-time signal is

$$X_d(z) = \sum_{i=1}^{n} \frac{c_i z}{z - e^{p_i T}}. \tag{7.5}$$

One finds that every pole p_i of the continuous-time signal is associated with a pole $p_{d,i}$ of the discrete-time signal equal to

$$p_{d,i} = e^{p_i T}. \tag{7.6}$$

Note that, because zeros are not transformed in the same manner, there can be pole/zero cancellations in the z-transform even if there are none in the Laplace transform. As a result, there may be fewer discrete-time poles than continuous-time poles.

Using the general formula of partial fraction expansion, the result can be extended to the general case with repeated poles. Therefore, a signal with rational transform in the s-domain always becomes a signal with a rational transform in the z-domain. The z-transform can be obtained relatively easily, and the poles of $X_d(z)$ are those of $X(s)$ transformed through $z = e^{sT}$.

Example 1: consider the signal

$$x(t) = e^{-at}u(t) \Leftrightarrow X(s) = \frac{1}{s+a}. \tag{7.7}$$

The discrete-time signal obtained by sampling is

$$x_d(k) = e^{-akT} = (e^{-aT})^k, \tag{7.8}$$

and has the transform

$$X_d(z) = \frac{z}{z - e^{-aT}}. \tag{7.9}$$

In other words, a signal with a first-order rational transform in the s-domain becomes a signal with a first-order rational transform in the z-domain. Note that the pole at $-a$ is transformed into $z = e^{-aT}$.

Example 2: this example shows that there may be a pole/zero cancellation in the formula for $X(z)$. Consider

$$x(t) = \cos(\omega_0 t) \Leftrightarrow X(s) = \frac{s}{s^2 + \omega_0^2}. \tag{7.10}$$

The discrete-time signal is

$$x_d(k) = \cos(\omega_0 T k) \Leftrightarrow X_d(z) = \frac{z^2 - \cos(\omega_0 T)z}{z^2 - 2\cos(\omega_0 T)z + 1}. \tag{7.11}$$

The transform has poles at $e^{j\omega_0 T}$ and $e^{-j\omega_0 T}$. It also has zeros at $z = 0$ and $z = \cos(\omega_0 T)$.

If $\omega_0 T = 2\pi$, $X_d(z)$ reduces to $X_d(z) - z/(z-1)$ and one pole is cancelled by a zero. Note that the transform is the same as the transform of a step signal. The two signals have the same transforms because, as shown in Fig. 7.2, the samples are identical. In general, the number of poles of $X_d(z)$ may be smaller than the number of poles of $X(s)$ (it cannot be larger, however), and it is not always possible to invert the sampling operation.

7.1.3 Transformation $z = e^{sT}$

The transformation $z = e^{sT}$ is at the core of the sampling operation, so that we analyze it in more detail. Since $e^{sT} = e^{\text{Re}(s)T}e^{j\,\text{Im}(s)T}$, we have that

$$|z| = e^{\text{Re}(s)T},$$
$$\angle z = \text{Im}(s)T. \tag{7.12}$$

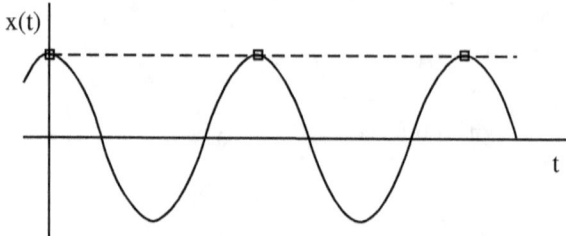

Figure 7.2: A cosine function and a step function sampled at a period equal to the period of the cosine function are identical signals

Note that the transformation $z = e^{sT}$ is not invertible, unless a restriction is placed on the variable s. Choose the portion of the s-plane with $-\frac{\pi}{T} < \text{Im}(s) \leqslant \frac{\pi}{T}$. With this restriction, the transformation is a bijection (one-to-one) and the inverse of the transformation is

$$s = \frac{1}{T}\ln(z) = \frac{1}{T}\ln(|z|) + j\,\frac{1}{T}\angle z. \qquad (7.13)$$

The mapping is such that

$$
\begin{array}{lll}
\text{Re}(s) \;<\; 0 & \text{if and only if} & |z| < 1 \\
\text{Re}(s) \;=\; 0 & \text{if and only if} & |z| = 1 \\
\text{Re}(s) \;>\; 0 & \text{if and only if} & |z| > 1
\end{array}
\qquad (7.14)
$$

The $j\omega$-axis of the s-plane maps to the unit circle of the z-plane, the open left half-plane maps to the inside of the unit circle, and the open right half-plane maps to the outside of the unit circle. This is shown in Fig. 7.3.

The equivalences of the following table are also worth noting.

s-plane	z-plane
$s = 0$	$z = 1$
$s = \pm j\frac{\pi}{T}$	$z = -1$
$s = +j\frac{\pi}{2T}$	$z = +j$
$s = -j\frac{\pi}{2T}$	$z = -j$
$\text{Re}(s) = -\infty$	$z = 0$
$s_2 = s_1^*$	$z_2 = z_1^*$

Note that a single z location is associated to a given s location, but the reverse is not true if all values of s are considered. Values of s outside the range $-\frac{\pi}{T} <$

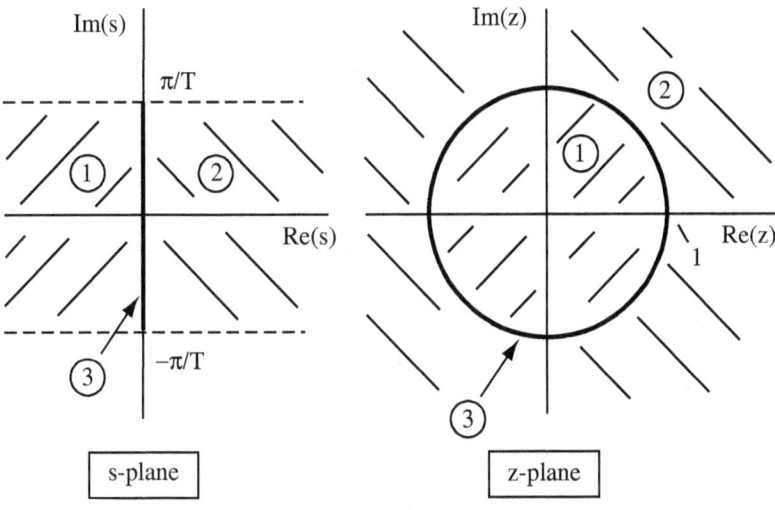

Figure 7.3: Mapping $z = e^{sT}$

$\text{Im}(s) \leq \frac{\pi}{T}$ map to the same values of z as some values of s within the range, with

$$s_2 = s_1 + j\frac{2\pi}{T} \Rightarrow z_2 = z_1. \qquad (7.15)$$

In other words, all complex numbers separated by a multiple of $j2\pi/T$ map to the same value of z. Note that $2\pi/T$ is equal to ω_s, the sampling frequency.

7.1.4 Transform of a sampled signal - General case

We now discuss the general relationship between the continuous-time and discrete-time transforms of a sampled signal. The transform $X(s)$ is not assumed to be rational.

Fact - Transform of a sampled signal

For $x(t)$ and $x_d(k)$ related by

$$x_d(k) = x(kT), \qquad (7.16)$$

the corresponding Laplace transform $X(s)$ and z-transform $X_d(z)$ are related by

$$X_d(z) = \left[\frac{1}{T} \sum_{n=-\infty}^{\infty} X\left(s - n\,j\frac{2\pi}{T} \right) \right]_{s=(1/T)\ln(z)}, \qquad (7.17)$$

or

$$[X_d(z)]_{z=e^{sT}} = \frac{1}{T} \sum_{n=-\infty}^{\infty} X\left(s - n\,j\frac{2\pi}{T}\right). \tag{7.18}$$

The proof of this fact is somewhat complicated and is left to the appendix at the end of the chapter.

The result involves an infinite sum of Laplace transforms, each shifted from the original one by a multiple of $j2\pi/T$, and the change of variable $z = e^{sT}$. An important observation is that the shift occurs along the direction of the imaginary axis by an amount of $2\pi/T$, which is exactly the shift that produces the same value of z in the transformation $z = e^{sT}$. This property implies that the transformation (7.18) gives the same result for $X_d(z)$ if values of s separated by $j2\pi/T$ are chosen. Hence, the non-invertibility of the transformation $z = e^{sT}$ is not an issue.

We found earlier that, for the continuous-time signal

$$x(t) = e^{-at}u(t) \Leftrightarrow X(s) = \frac{1}{s+a}, \tag{7.19}$$

the z-transform of the sampled signal was

$$X_d(z) = \frac{z}{z - e^{-aT}}. \tag{7.20}$$

In contrast, the general result gives

$$X_d(z) = \left[\frac{1}{T} \sum_{n=-\infty}^{\infty} \frac{1}{s + a - n\,j\frac{2\pi}{T}}\right]_{s=(1/T)\ln(z)}. \tag{7.21}$$

Therefore, the series (7.21) must be equal to the analytic form (7.20). However, an analytic form of the infinite series cannot be found, in general.

7.1.5   Transform of a sampled signal in the frequency domain

Consider the Fourier transform (FT) of the continuous-time signal $x(t)$, which we will denote $\bar{X}(\omega)$, and the discrete-time Fourier transform (DTFT) of the discrete-time signal $x_d(k)$, which we will denote $\bar{X}_d(\Omega)$, with

$$\bar{X}(\omega) = \int_{-\infty}^{\infty} x(t)e^{-j\omega t}\,dt \qquad \text{(Fourier transform)}$$

$$\bar{X}_d(\Omega) = \sum_{k=-\infty}^{\infty} x_d(k)e^{-j\Omega k} \qquad \text{(Discrete-time Fourier transform)}. \tag{7.22}$$

Because the frequency-domain transforms are defined as bilateral transforms, while our definitions of Laplace and z-transforms are unilateral, we will assume that the signals are zero for negative time.

Assuming that the Fourier transforms exist, we have that

$$\begin{aligned}
\bar{X}(\omega) &= [X(s)]_{s=j\omega}, \\
\bar{X}_d(\Omega) &= [X_d(z)]_{z=e^{j\Omega}}.
\end{aligned} \tag{7.23}$$

The relationships between the s-plane and the ω variable, and between the z-plane and the Ω variable, are shown in Fig. 7.4.

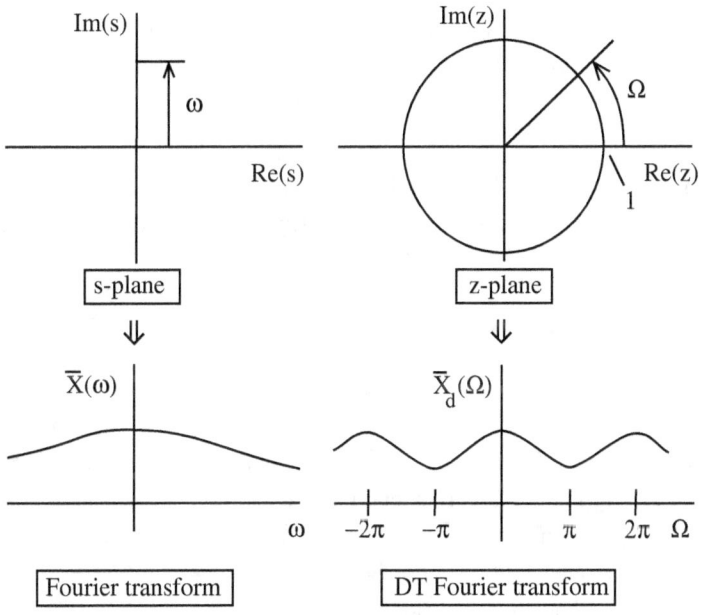

Figure 7.4: From the Laplace and z-transforms to the Fourier and DT Fourier transforms

In the frequency domain, the transformation $z = e^{sT}$ becomes

$$e^{j\Omega} = z = e^{sT} = e^{j\omega T}, \tag{7.24}$$

or simply

$$\Omega = \omega T. \tag{7.25}$$

With (7.25), the formula (7.17) becomes

$$\bar{X}_d(\Omega) = \frac{1}{T} \sum_{n=-\infty}^{\infty} \left[\bar{X}\left(\omega - n\frac{2\pi}{T}\right) \right]_{\omega=\Omega/T}. \tag{7.26}$$

This result allows one to calculate the DTFT of the discrete-time signal $x_d(k)$, knowing the FT of the continuous-time signal $x(t)$. The transformation in the frequency domain is composed of two steps:

- a rescaling of the frequency axis, so that $\Omega = \omega T$. In particular, the sampling frequency $\omega_s = 2\pi/T$ is mapped to $\Omega = 2\pi$. For example, a 100 Hz signal sampled at 1 kHz is mapped to $\pi/5$.

- an infinite sum of the transform shifted by multiples of $2\pi/T$ in the ω space, or 2π in the Ω space.

- a scaling of the amplitude of the transform by a factor $1/T$.

Note that, while the result was derived assuming that the signals were zero for negative time, (7.26) is valid for arbitrary signals, provided that their Fourier transforms exist.

7.1.6   Aliasing

Fig. 7.5 shows the transform of a continuous-time signal $x(t)$, with $\bar{X}(\omega) = 0$ for $|\omega| > \omega_B$. *For simplicity of presentation, $\bar{X}(\omega)$ is taken to be a real function of ω, but it is a complex function, in general.* Assume that the signal is sampled with a sampling period T, corresponding to a sampling frequency ω_s. The top figures show the result that is obtained when $\omega_B < \omega_s/2 = \pi/T$, so that the discrete-time frequency $\Omega_B = \omega_B T$ is less than π. The figures on the bottom show the result when this condition is not satisfied.

   In the first case, only the original copy of the transform contributes to the discrete-time transform in the $-\pi$ to π range. In the second case, there is interference between the original transform and its copies. This situation is called *aliasing*. When there is aliasing, a frequency component may be observed in the discrete-time transform which is the image of a higher frequency in the original signal.

   Fig. 7.6 demonstrates this phenomenon in the time domain. A signal with frequency $\omega_0 = \pi/4$ (period of 8 seconds) is sampled at $\omega_s = \pi/3$ (period of 6 seconds). In the discrete-time domain, a frequency of $\Omega = 2\pi - \omega_0 T = 2\pi - 3\pi/2 = \pi/2$ is obtained, which is the same as would have been obtained with a

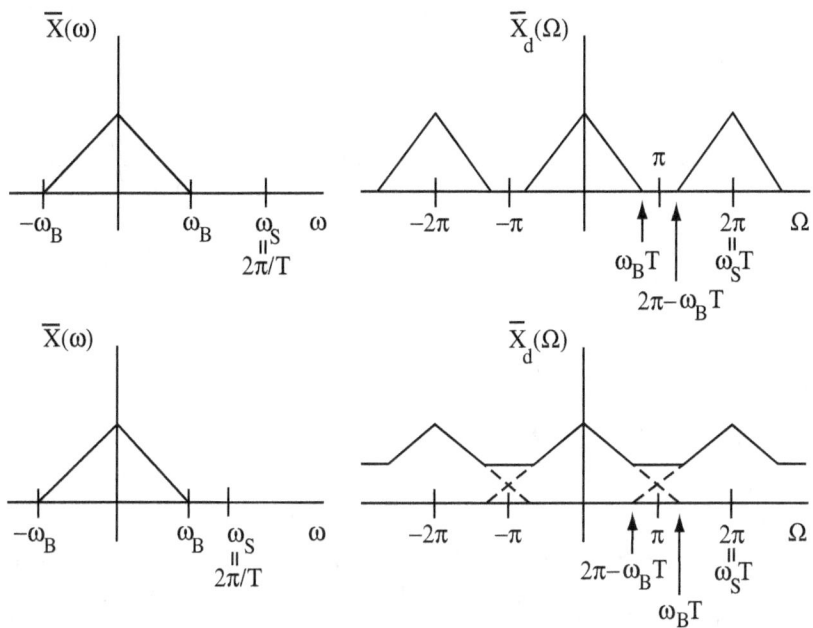

Figure 7.5: Transformation of a band-limited transform without aliasing (top) and with aliasing (bottom)

frequency $\Omega = \omega_0 T = \pi/2$ or $\omega_0 = \pi/12$ (period of 24 seconds). The two signals that yield the same samples are shown on the figure. A similar phenomenon was also observed in Fig. 7.2.

Numerical example: let the sampling frequency be 1000 Hz. 500 Hz is the maximum frequency that can be sampled without aliasing. A frequency of 500 Hz maps to π in the discrete-time frequency domain. A frequency of 100 Hz maps to $\pi/5$, and 250 Hz maps to $\pi/2$. A frequency of 600 Hz maps to $4\pi/5$, which is the same frequency as 400 Hz. Similary, 800 Hz maps to the same frequency as 200 Hz, and 1000 Hz is the same as a DC signal. Higher frequencies make confusion worse. For example, 100 Hz is indistinguishable from 900 Hz, 1100 Hz, 1900 Hz, 2100 Hz, ...(in general, a frequency f cannot be distinguished from $nf_s \pm f$, where f_s is the sampling frequency and $n = 1, 2, ...$).

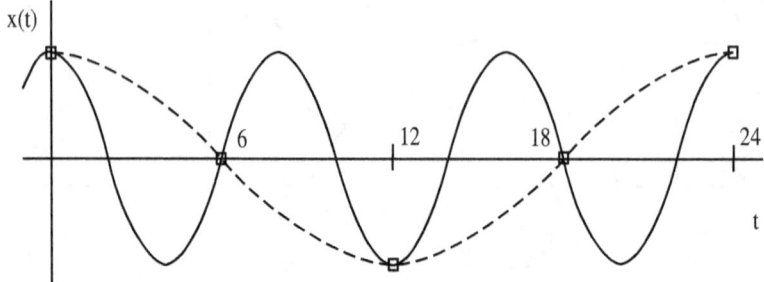

Figure 7.6: Two signals yielding the same samples

7.1.7 Avoiding aliasing

When there is aliasing, it is impossible to recover unambiguously a signal from its samples. In order to avoid such situation, *the sampling frequency must be at least twice the highest frequency present in the signal to be sampled.* The minimum sampling frequency for a given signal, *i.e.*, $\omega_s = 2\omega_B$, is called *the Nyquist rate.* Since real signals are rarely, if ever, bandlimited, one must generally choose a frequency range of interest and a sampling frequency equal to at least twice the upper bound. Aliasing is then prevented by use of an *anti-aliasing filter.* Such a filter is applied to the signal before discretization. An ideal anti-aliasing filter is a lowpass filter whose gain is 1 for $|\omega| < \omega_s/2$ and 0 otherwise. The phase should be zero for $|\omega| < \omega_s/2$. Such a filter is not realizable and an approximation must be implemented. In practice, the bandwidth of the filter is typically specified slightly lower than half the sampling frequency in order to account for the transition between the passing band and the stopping band of any practical anti-aliasing filter $\bar{F}(\omega)$ (see Fig. 7.7).

If the conditions are satisfied so that aliasing is avoided, the relationship between the DTFT of the discrete-time signal and the FT of the continuous-time signal is

$$\bar{X}_d(\Omega) = \frac{1}{T} \left[\bar{X}(\omega) \right]_{\omega=\Omega/T} \qquad \text{for } -\pi < \Omega \leqslant \pi. \qquad (7.27)$$

The transformation simply amounts to a rescaling of the frequency variable and of the transform. Shifted copies of the continuous-time transform do not affect the discrete-time transform. Note that (7.27) is a special case of

$$\left[X_d(z) \right]_{z=e^{sT}} = \frac{1}{T} X(s) \qquad \text{or} \qquad X_d(z) = \left[\frac{1}{T} X(s) \right]_{s=(1/T)\ln(z)} . \qquad (7.28)$$

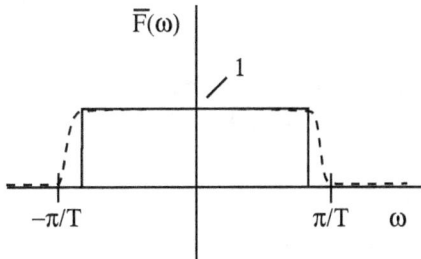

Figure 7.7: Anti-aliasing filter

7.2 Conversion discrete-time signal to continuous-time signal

7.2.1 Definition of reconstruction

We now consider the reconstruction of a continuous-time signal $x(t)$ from a discrete-time signal $x_d(k)$ by

$$x(t) = x_d(k) \qquad \text{for } t \in [kT, (k+1)T). \tag{7.29}$$

The transformation is shown in Fig. 7.8. The operation is the one commonly performed in digital-to-analog (D/A) converters, and is usually referred to as a *zero-order hold* (ZOH). More sophisticated converters exist that interpolate the values of $x(t)$ between the time instants. A linear interpolator would be called a *first-order hold*. However, the zero order hold is by far the most commonly used.

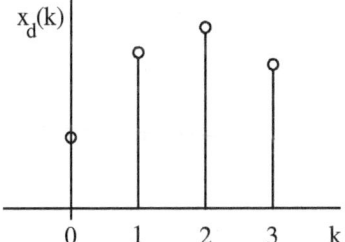 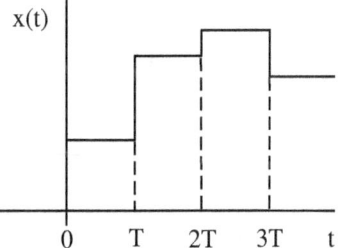

Figure 7.8: Conversion of a discrete-time signal to a continuous-time signal

7.2.2   Transform of a reconstructed signal

Again, a natural problem is to relate $X(s)$, the Laplace transform of $x(t)$, to $X_d(z)$, the z-transform of $x_d(k)$. Here, the result is simpler.

Fact - Transform of a reconstructed signal

For $x(t)$ and $x_d(k)$ related by (7.29), the Laplace transform $X(s)$ and the z-transform $X_d(z)$ are related by

$$
\begin{aligned}
X(s) &= [X_d(z)]_{z=e^{sT}} \; \frac{1 - e^{-sT}}{s} \\
&= T \; [X_d(z)]_{z=e^{sT}} \; \frac{1 - e^{-sT}}{sT}.
\end{aligned}
\tag{7.30}
$$

The proof is not difficult in this case, but is left to the appendix at the end of the chapter.

Example 1: consider a step $x_d(k) = u(k)$. Then, $x(t) = u(t)$ is also a step (though in continuous-time instead of discrete-time). Applying the formula with $X_d(z) = z/(z - 1)$ yields $X(s) = 1/s$, as expected.

Example 2: consider $x_d(k) = \delta(k)$, the discrete-time impulse. Now $x(t)$ is *not* a continuous-time impulse, but a pulse of duration T, *i.e.*, a function equal to 1 for $0 \leqslant t < T$ and zero otherwise (see Fig. 7.9). As expected, $X_d(z) = 1$ yields $X(s) = (1 - e^{-sT})/s$ (recall that the Laplace transform associated with a pure delay T is e^{-sT}). Note that $X(s)$ is *not* a rational function of s, although $X_d(z)$ is a rational function of z. Indeed, the discrete to continuous conversion does *not* preserve the rational nature of transforms, in general.

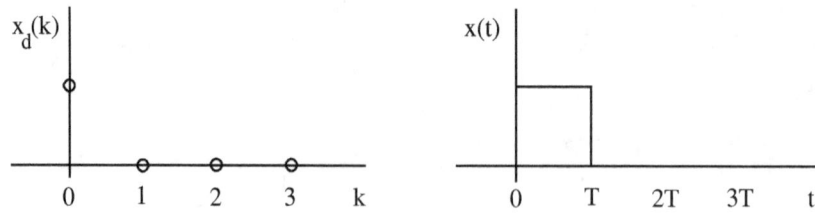

Figure 7.9: Conversion of a discrete-time impulse to continuous-time

7.2.3   Transform of a reconstructed signal in the frequency domain

Again, we define $\bar{X}(\omega)$ as the Fourier transform of $x(t)$, and $\bar{X}_d(\Omega)$ as the discrete-time Fourier transform of $x_d(k)$. The transformation $z = e^{sT}$ becomes

$\Omega = \omega T$, so that the formula (7.30) becomes

$$\bar{X}(\omega) = T \ \left[\bar{X}_d(\Omega)\right]_{\Omega=\omega T} \left(\frac{1 - e^{-j\omega T}}{j\omega T}\right). \tag{7.31}$$

The transformation is composed of three steps:

- a rescaling of the frequency axis such that $\Omega = \omega T$ or $\omega = \Omega/T$.

- a filtering of the resulting signal by $(1 - e^{-j\omega T})/j\omega T$.

- a scaling of the magnitude of the transform by T.

Zero-order hold transfer function
The transfer function

$$H(s) = \frac{1 - e^{-sT}}{sT} \tag{7.32}$$

is referred to as the transfer function of the zero-order hold. In the frequency domain, it can be expressed as

$$\begin{aligned}
\bar{H}(\omega) &= \frac{1 - e^{-j\omega T}}{j\omega T} = e^{-j\omega T/2} \frac{e^{j\omega T/2} - e^{-j\omega T/2}}{j\omega T} \\
&= e^{-j\omega T/2} \frac{2 \sin(\omega T/2)}{\omega T} \\
&= e^{-j\omega T/2} \ \mathrm{sinc}\,(\omega T/2), \tag{7.33}
\end{aligned}$$

where $\mathrm{sinc}(x) \triangleq \sin(x)/x$. The magnitude of the frequency response is $|\mathrm{sinc}\,(\omega T/2)|$. The phase is $-\omega T/2$, plus π when $\mathrm{sinc}(\omega T/2)$ is negative. The magnitude and phase are shown in Fig. 7.10. Note that, for frequencies below $2\pi/T$, the phase lag is equal to the phase lag produced by a time delay of $T/2$.

Example: the effects of the discrete to continuous conversion may be studied further by considering a discrete-time sinusoid $x_d(k) = \cos(\Omega_0 k)$. The transform is a pair of impulses at $\pm\Omega_0$, with magnitude $1/2$, plus the copies shifted by all the multiples of 2π. The resulting transform of the continuous-time signal is shown on Fig. 7.11. Interestingly, one has that

$$\int_{-\infty}^{\infty} T \ \left[\frac{1}{2}\delta(\Omega - \Omega_0)\right]_{\Omega=\omega T} d\omega = \int_{-\infty}^{\infty} \frac{1}{2}\delta(\Omega - \Omega_0)d\Omega, \tag{7.34}$$

so that the scaling of the axes cancels the factor T in the formula and the magnitude of the impulses remains $1/2$ in continuous-time (except for the slight reduction in magnitude due to the zero-order hold).

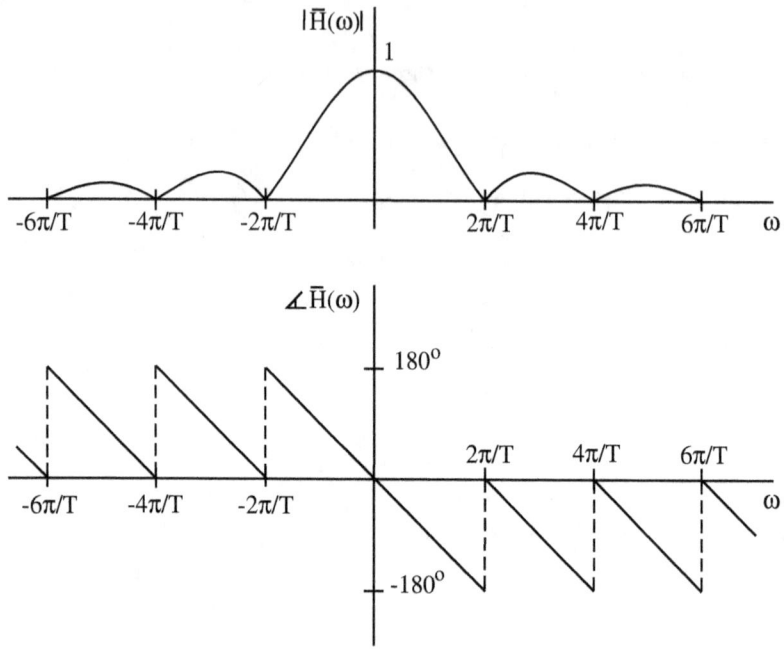

Figure 7.10: Frequency response of a zero-order hold

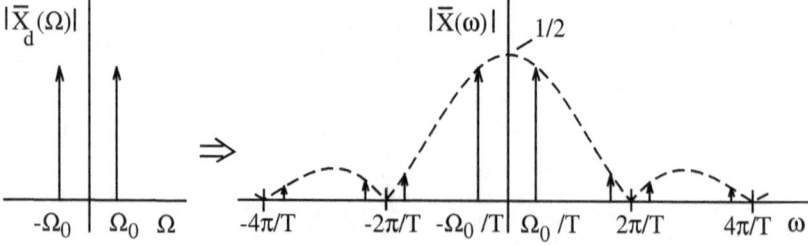

Figure 7.11: Transform of a sinusoid reconstructed through a zero-order hold

The transform of the discrete-time signal is *not* that of a continuous-time sinusoid. Indeed, the signal is shown in Fig. 7.12, and is discontinuous. From knowledge of Fourier series, one should expect a number of high-frequency components associated with the sharp transitions at the sampling times. The two large components of the transform correspond to the fundamental of the signal, which has a magnitude slightly less than the original signal, and a phase shift corresponding to a time delay of $T/2$. This phase can be anticipated from the shape of the reconstructed signal. Interestingly, the reconstructed signal is *not* periodic, unless the sampling frequency is a multiple of the signal frequency, as in Fig. 7.12. The visible result on an oscilloscope is that the stepwise component of the waveform shifts continuously with respect to the fundamental.

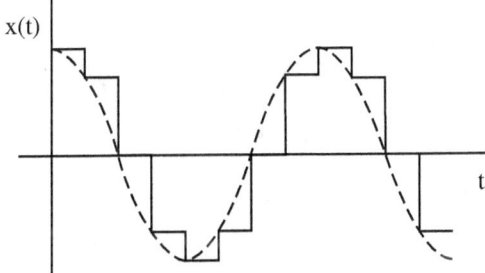

Figure 7.12: Sinusoid reconstructed through zero-order hold

Numerical example: let $\Omega_0 = \pi/4$, which yields the signal shown in Fig. 7.12. Assume that the sampling frequency is $f_s = 1000$ Hz (or $T = 1$ ms). The continuous-time signal is of the form

$$x_1(t) = M_1 \cos(\omega_1 t + \phi_1) + M_2 \cos(\omega_2 t + \phi_2) + ... \qquad (7.35)$$

The frequency of the first component is $\omega_1 = \Omega_0/T = \pi/(4T)$, or $f_1 = \omega_1/(2\pi) = 1/(8T) = f_s/8$ or 125 Hz. The magnitude and phase of the first component are

$$M_1 = \left| \frac{\sin(\omega_1 T/2)}{\omega_1 T/2} \right| = \frac{\sin(\pi/8)}{\pi/8} = 0.975$$
$$\phi_1 = -\omega_1 T/2 = -\pi/8 = -22.5°. \qquad (7.36)$$

For the second component, $\omega_2 = 2\pi/T - \Omega_0/T = 7\pi/(8T)$. The frequency of this component is $f_2 = \omega_2/(2\pi) = 7f_s/8$ or 875 Hz. The magnitude and phase of the second component are

$$M_2 = \left| \frac{\sin(\omega_2 T/2)}{\omega_2 T/2} \right| = \frac{\sin(7\pi/8)}{7\pi/8} = 0.139$$

$$\phi_2 = -\omega_2 T/2 = -7\pi/8 = -157.5°. \qquad (7.37)$$

Similarly, the third component has frequency 1125 Hz, magnitude 0.108, and phase $-22.5°$. Overall, one finds that the reconstructed signal has a fundamental component at the desired frequency, but with a slightly lower magnitude (2.5% smaller) and a significant phase delay (22.5°). Additional components are present at higher frequencies with magnitudes of about 10% of the fundamental. Generally, these are *not* harmonic frequencies, but rather multiples of the sampling frequency plus or minus the fundamental signal frequency. Parasitic effects are reduced if the ratio of the fundamental frequency to the sampling frequency decreases.

7.3   Conversion continuous-time system to discrete-time system

7.3.1   Equivalent discrete-time system

The need to find a discrete-time equivalent to a continuous-time system arises in control applications, such as shown in Fig. 7.13. A plant is controlled by a computer-based system, such that the control input $x(t)$ is generated by a D/A converter, the plant output $y(t)$ is sampled by an A/D converter, and the discrete-time control input is calculated by a computer (microprocessor, or other). Two approaches are possible: the first consists in designing a continuous-time control law for the plant, and then finding a discrete-time equivalent for implementation. This approach is discussed later. The second approach consists in finding a discrete-time equivalent to the plant, that is, a description for the transformation from $x_d(k)$ to $y_d(k)$. This approach is discussed now.

An interesting result is that the system from $x_d(k)$ to $y_d(k)$ *is* a linear time-invariant system. Its transfer function is such that the step response of the discrete-time system matches the samples of the step response of the continuous-time system (see Fig. 7.14). Surprisingly, the result holds true without the assumption of ideal anti-aliasing or post-sampling filters (although these filters are nevertheless useful in practice).

Example: consider a first-order system

$$P(s) = \frac{1}{s+1}. \qquad (7.38)$$

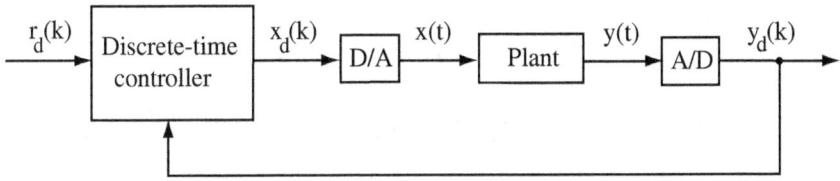

Figure 7.13: Digital control application

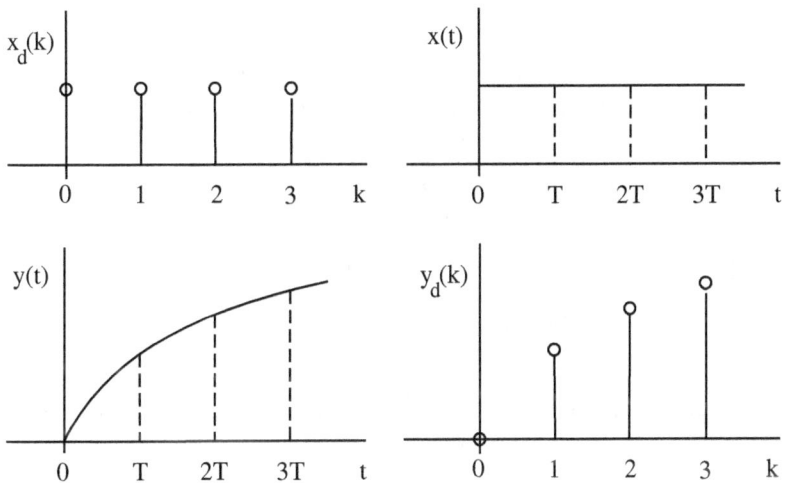

Figure 7.14: Step response matching

If a discrete-time step input $x_d(k)$ is applied to the D/A, the result is a continuous-time step input $x(t)$ applied to the plant. The step response is

$$Y(s) = \frac{1}{(s+1)s} = \frac{1}{s} - \frac{1}{s+1} \quad \Leftrightarrow \quad y(t) = 1 - e^{-t}. \tag{7.39}$$

The sampled output of the step response is

$$y_d(k) = 1 - e^{-kT}$$
$$\Leftrightarrow \quad Y_d(z) = \frac{z}{z-1} - \frac{z}{z-e^{-T}} = \frac{z(1-e^{-T})}{(z-1)(z-e^{-T})}. \tag{7.40}$$

On the other hand, the step response of the equivalent discrete-time system is

$$Y_d(z) = P_d(z)\frac{z}{z-1}. \tag{7.41}$$

We conclude that

$$P_d(z) = \frac{1 - e^{-T}}{z - e^{-T}}. \tag{7.42}$$

$P_d(z)$ is usually called the *step response equivalent* or *zero-order hold equivalent* of $P(s)$.

Although only the step response of the discrete-time system was shown to match the step response of the sampled-data system, it is not hard to show that the responses of both systems are identical for any input signal $x_d(k)$. Indeed, any $x_d(k)$ may be viewed as the superposition of shifted step signals, and linear time-invariance implies the result.

The procedure to obtain $P_d(z)$ can also be extended to arbitrary linear systems with rational transforms, and is similar to the procedure associated with the discretization of a signal with rational transform. The notation becomes complicated for repeated poles, so we assume that $P(s) = N(s)/D(s)$ is rational and strictly proper, has non-repeated poles, and has no pole at $s = 0$. The step response of the continuous-time system is given by

$$Y(s) = P(s)\frac{1}{s} = P(0)\frac{1}{s} + \sum_{i=1}^{n}\frac{c_i}{s - p_i} \quad \Leftrightarrow \quad y(t) = P(0) + \sum_{i=1}^{n}c_i e^{p_i t}, \tag{7.43}$$

where p_i are the poles of the transfer function $P(s)$ and c_i are the coefficients of the partial fraction expansion, with

$$c_i = \left[\frac{s - p_i}{s}P(s)\right]_{s=p_i}. \tag{7.44}$$

If we sample this signal every T seconds, the resulting discrete-time signal is given by

$$y_d(k) = P(0) + \sum_{i=1}^{n}c_i e^{p_i k T} \quad \Leftrightarrow \quad Y_d(z) = P(0)\frac{z}{z - 1} + \sum_{i=1}^{n}c_i \frac{z}{z - e^{p_i T}}. \tag{7.45}$$

Using (7.41), the transfer function of the discrete-time system is given by

$$P_d(z) = P(0) + (z - 1)\sum_{i=1}^{n}c_i\frac{1}{z - e^{p_i T}}. \tag{7.46}$$

The poles p_i of the transfer function $P(s)$ are mapped to poles $e^{p_i T}$ of $P_d(z)$. Therefore,

$$P(s) \text{ is rational of order } n \quad \Rightarrow \quad P_d(z) \text{ is rational of order at most } n$$
$$P(s) \text{ is asymptotically stable} \quad \Rightarrow \quad P_d(z) \text{ is asymptotically stable. (7.47)}$$

While the poles are mapped through $z = e^{sT}$, the zeros are not necessarily mapped in the same manner. Therefore, pole/zero cancellations may cause $P_d(z)$ to have fewer poles than $P(s)$, and the order of the transfer function may be reduced. The frequency responses of the continuous-time and discrete-time systems are also not obviously related. However, $P_d(1) = P(0)$, so that the DC gains of the two systems are identical. For T sufficiently small, one can also show that

$$[P_d(z)]_{z=e^{sT}} \simeq P(s) \qquad \text{for } |sT| \ll 1, \tag{7.48}$$

so that the step response equivalent and the transformation $z = e^{sT}$ give the same result for low frequencies.

7.3.2   Discrete-time controller for continuous-time plant

Given a discrete-time equivalent $P_d(z)$ to the continuous-time plant $P(s)$, Fig. 7.15 shows how a discrete-time control algorithm $C_d(z)$ can be analyzed in the z-domain. Root-locus methods may be used, for example, to place the poles of the discrete-time system. Desirable locations must be considered in the z-plane, instead of the s-plane (see Fig. 6.22).

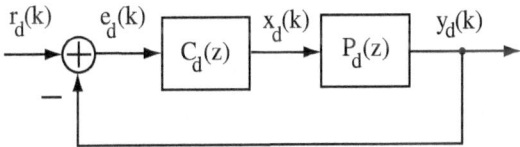

Figure 7.15: Direct design of digital controllers in the z-domain

For example, consider the continuous-time system

$$P(s) = \frac{1}{s+1}, \tag{7.49}$$

with $C(s) = g$. The continuous-time controller gives a stable closed-loop system for all g (the root-locus shows a pole moving along the real axis in the negative direction).

The equivalent discrete-time system is

$$P_d(z) = \frac{1 - e^{-T}}{z - e^{-T}}. \tag{7.50}$$

A discrete-time controller $C_d(z) = g$ gives a closed-loop pole at

$$z = e^{-T} - k(1 - e^{-T}).$$ (7.51)

As in continuous-time, the pole moves along the real axis in the negative direction. The pole becomes unstable when it reaches $z = -1$ for

$$g_{max} = \frac{1 + e^{-T}}{1 - e^{-T}}.$$ (7.52)

In fact, there is no benefit in pushing the pole further than $z = 0$, so that a practical limit for the gain is

$$g_0 = \frac{e^{-T}}{1 - e^{-T}}.$$ (7.53)

The gain g_0 results in the transfer function

$$P_{CL}(z) = \frac{e^{-T}}{z}.$$ (7.54)

which is a one-step delay with a gain e^{-T}. Typically, such response (called *deadbeat response*) requires large input signals and is sensitive to noise and unmodelled dynamics, so that the feedback gain will be set much below g_{max}.

7.4 Conversion discrete-time system to continuous-time system

7.4.1 Equivalent continuous-time system

We now turn to the problem of converting a discrete-time system to a continuous-time system. This problem arises, in particular, in the digital filtering application shown in Fig. 7.16. The objective is to filter the signal $x(t)$ to remove some undesirable frequency components. Instead of implementing a continuous-time filter, a digital processing system is used so that the signal $x(t)$ is sampled through an A/D converter, processed by a computing device, and transformed back to a continuous-time signal $y(t)$ through a D/A converter. Another situation is where a compensator is designed in continuous-time, and a discrete-time implementation is sought. The computing element operates in discrete-time, and implements a linear time-invariant system with transfer function $F_d(z)$. An important question is: what is the relationship between the transfer function $F_d(z)$ and the transformation from $x(t)$ to $y(t)$?

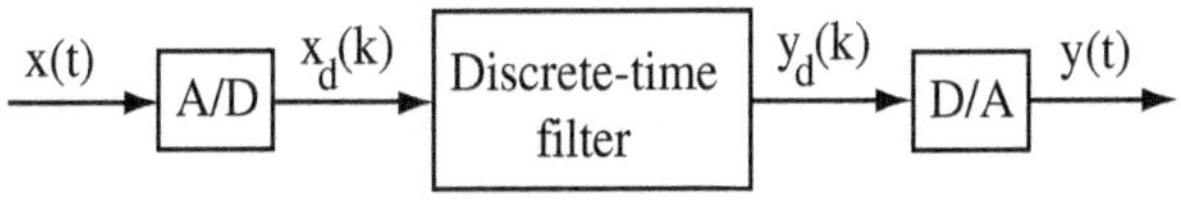

Figure 7.16: Digital filtering application

As opposed to the (reverse) problem discussed in the previous section, the system shown in Fig. 7.16 is generally *not* time-invariant, even if the discrete-time system is. In particular the response of the system depends on how the sampling instants are synchronized with the input signal (consider for example the effect of a delay of sampling times in Fig. 7.2). Using previous results, we have, in general

$$X_d(z) = \left[\frac{1}{T} \sum_{k=-\infty}^{k=\infty} X\left(s - k\, j\frac{2\pi}{T} \right) \right]_{s=(1/T)\ln(z)}, \qquad (7.55)$$

so that

$$Y_d(z) = F_d(z) \left[\frac{1}{T} \sum_{k=-\infty}^{k=\infty} X\left(s - k\, j\frac{2\pi}{T} \right) \right]_{s=(1/T)\ln(z)}, \qquad (7.56)$$

and

$$\begin{aligned}
Y(s) &= T\, [Y_d(z)]_{z=e^{sT}}\, \frac{1 - e^{-sT}}{sT} \\
&= [F_d(z)]_{z=e^{sT}}\, \frac{1 - e^{-sT}}{sT} \sum_{k=-\infty}^{k=\infty} X\left(s - k\, j\frac{2\pi}{T} \right).
\end{aligned} \qquad (7.57)$$

This transformation relates the Laplace transforms of the input and output of the system. Unfortunately, the transformation cannot be put into the form $Y(s) = F(s)X(s)$ due to the change of variables $z = e^{sT}$ and due to the infinite sum.

If the aliasing effects can be neglected, only the term $k = 0$ is retained, and we have the approximation

$$Y(s) = [F_d(z)]_{z=e^{sT}}\, \frac{1 - e^{-sT}}{sT} X(s). \qquad (7.58)$$

Thus, the transformation can be approximately represented by a transfer function $F(s)$, with

$$F(s) = [F_d(z)]_{z=e^{sT}}\, \frac{1 - e^{-sT}}{sT} \qquad \text{(aliasing effects neglected)}. \qquad (7.59)$$

Further, if the zero-order hold effects are neglected

$$F(s) = [F_d(z)]_{z=e^{sT}} \qquad \text{(aliasing and ZOH effects neglected)}.$$

(7.60)

Even with these assumptions, a rational transfer function $F_d(z)$ does not yield a rational transfer function $F(s)$.

7.4.2 Equivalent system in the frequency domain

In the frequency domain, the corresponding relationships are

$$\bar{Y}(\omega) = [\bar{F}_d(\Omega)]_{\Omega=\omega T} \frac{1 - e^{-j\omega T}}{j\omega T} \sum_{k=-\infty}^{k=\infty} \bar{X}\left(\omega - k\frac{2\pi}{T}\right).$$

(7.61)

If antialiasing conditions are satisfied,

$$\bar{X}(\omega) = 0 \qquad \text{for } |\omega| > \frac{\pi}{T},$$

(7.62)

and if zero-order hold effects are also neglected, (7.61) reduces to

$$\begin{aligned} \bar{Y}(\omega) &= [\bar{F}_d(\Omega)]_{\Omega=\omega T} \, \bar{X}(\omega) & \text{for } |\omega| < \frac{\pi}{T} \\ &= 0 & \text{otherwise.} \end{aligned}$$

(7.63)

In other words, the overall system with ideal anti-aliasing and post-sampling filters is equivalent to a continuous-time filter

$$\bar{F}(\omega) = [\bar{F}_d(\Omega)]_{\Omega=\omega T} \qquad \text{for } |\omega| < \frac{\pi}{T}.$$

(7.64)

(7.64) is the equivalent, for Fourier transforms, of (7.60). Under these assumptions, the equivalence between the discrete-time filter and the continuous-time filter is represented on Fig. 7.17 for a discrete-time low-pass filter of bandwidth Ω_B.

7.4.3 Delay of a low-pass filter

In continuous-time, the delay of a low-pass filter was computed using formulas (5.27) and (5.29). Equivalent formulas can be derived for a discrete-time filter as well. In discrete-time, the low frequency behavior of a transfer function is described by the value of $F(z)$ in the vicinity of $z = 1$. The formula equivalent to (5.27) in discrete-time is

$$n_d = \lim_{\Delta z \to 0} \frac{F(1) - F(1 + \Delta z)}{\Delta z \, F(1)},$$

(7.65)

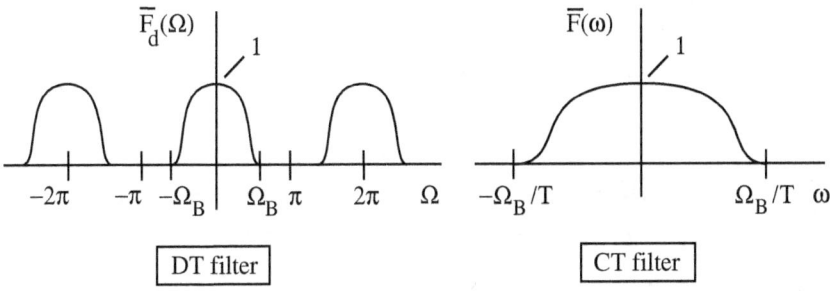

Figure 7.17: Discrete-time filter and equivalent continuous-time filter

where n_d is the low-frequency delay measured in samples. n_d is not necessarily an integer. For example, consider

$$F(z) = \frac{b}{z^2(z-a)}. \tag{7.66}$$

The formula gives

$$
\begin{aligned}
n_d &= \lim_{\Delta z \to 0} \frac{b(1+\Delta z)^2(1+\Delta z - a) - b(1-a)}{\Delta z\, b(1+\Delta z)^2(1+\Delta z - a)} \\
&= \frac{3-2a}{1-a},
\end{aligned}
\tag{7.67}
$$

or

$$n_d = 2 + \frac{1}{1-a}. \tag{7.68}$$

It can be checked that the first term is the delay computed for $1/z^2$ alone, while the second term is the delay resulting from the low-pass filter $1/(z-a)$. $1/(1-a)$ can be much larger than 2 if a is close to 1 (for example, $a = 0.99$ gives 100).

In general, for a rational transfer function

$$F(z) = \frac{b_{n-1}z^{n-1} + \cdots + b_1 z + b_0}{z^n + a_{n-1}z^{n-1} + \cdots + a_1 z + a_0} \tag{7.69}$$

the time delay is

$$n_d = \frac{n + (n-1)a_{n-1} + \cdots + 2a_2 + a_1}{1 + a_{n-1} + \cdots + a_1 + a_0} - \frac{(n-1)b_{n-1} + \cdots + 2b_2 + b_1}{b_{n-1} + \cdots + b_1 + b_0}. \tag{7.70}$$

This formula is the equivalent of (5.29) obtained in continuous-time.

If the discrete-time filter is implemented as in Fig. 7.16, the time delay (in seconds) associated with the filter may increase by $T/2$ due to the zero-order hold and another T due to the fact the output computed by the discrete-time filter is normally applied only at the next time instant. If the filter is inserted in a feedback system, the delay margin must be sufficient to accommodate the filter's delay.

7.5   Discrete-time design to approximate a continuous-time system

This problem is, in some ways, the combination of the two previous problems. Starting from a continuous-time system $F_c(s)$ (which, in a control application would be a compensator $C(s)$), find $F_d(z)$ such that the continuous-time system corresponding to $F_d(z)$ in Fig. 7.16 approximates $F(s)$, *i.e.*, $F(s) \simeq F_c(s)$. A simple answer would be to choose $F_d(z)$ such that $[F_d(z)]_{z=e^{sT}} = F_c(s)$, since the transformation is known to give $F(s) = F_c(s)$ (if aliasing and zero-order hold effects neglected). However, the transformation $z = e^{sT}$ does not preserve the rational nature of a transfer function, so that implementation of the discrete-time transfer function would not be possible as a difference equation, even if the continuous-time transfer function was rational. Therefore, further approximations are commonly used.

Euler approximation

The Euler approximation consists in approximating

$$\frac{dx}{dt} \simeq \frac{x(t+T) - x(t)}{T} \qquad \text{for } T \text{ small,} \tag{7.71}$$

which is equivalent to the following transformation

$$s = \frac{z-1}{T} \qquad \text{or} \qquad z = 1 + sT. \tag{7.72}$$

For example, the PID controller (4.35) becomes

$$C_d(z) = k_P + k_I \frac{T}{z-1} + k_D \frac{a(z-1)}{z-1+aT}. \tag{7.73}$$

Note that this transformation is an approximation of $z = e^{sT}$ for T small. It preserves the rational nature and order of the transfer function with

$$F_d(z) = [F_c(s)]_{s=(z-1)/T} . \tag{7.74}$$

However, stability may not be preserved. Continuous-time poles must be located inside the circle of radius $1/T$ and with center $-1/T$ for the discrete-time system to be stable. Nevertheless, for T sufficiently small, stability will be obtained. This simple method is often adequate, but requires caution.

Bilinear transformation (Tustin's method)

A better rational approximation of the transformation $z = e^{sT}$ is the *bilinear transformation*

$$z = e^{sT} = \frac{e^{sT/2}}{e^{-sT/2}} \simeq \frac{1 + sT/2}{1 - sT/2}. \tag{7.75}$$

The transformation is invertible, since

$$z\left(1 - \frac{sT}{2}\right) = 1 + \frac{sT}{2} \Rightarrow s = \frac{2}{T}\frac{z-1}{z+1}. \tag{7.76}$$

Given $F_c(s)$ with desirable frequency-domain properties, one lets

$$F_d(z) = [F_c(s)]_{s = \frac{2}{T}\frac{z-1}{z+1}}. \tag{7.77}$$

It is not difficult to verify that

$$F_c(s) \text{ is rational of order } n \quad \Leftrightarrow \quad F_d(z) \text{ is rational of order } n. \tag{7.78}$$

Further, it turns out that the stability regions of the s and z planes are mapped exactly to each other, that is

$$\text{Re}(s) < 0 \quad \Leftrightarrow \quad |z| > 1. \tag{7.79}$$

Therefore,

$$F_c(s) \text{ stable} \quad \Leftrightarrow \quad F_d(z) \text{ stable.} \tag{7.80}$$

From the properties of the conversion from discrete-time system to continuous-time system, we also know that

$$F_d(z) \text{ stable} \quad \Leftrightarrow \quad F(s) \text{ stable.} \tag{7.81}$$

Impulse response matching

The approach consists in matching the impulse response of the discrete-time system $f_d(k)$ to the impulse response of the continuous-time system $f_c(t)$. Recalling the results regarding the conversion of a continuous-time signal to a discrete-time signal, we have that

$$F_c(s) = \frac{N(s)}{D(s)} = \sum_{i=1}^{n} \frac{c_i}{s - p_i} \quad \Leftrightarrow \quad F_d(z) = \sum_{i=1}^{n} c_i \frac{z}{z - e^{p_i T}}, \tag{7.82}$$

where p_i are the poles of $F_c(s)$ and c_i are the coefficients of the partial fraction expansion (for simplicity, it is assumed that $F_c(s) = N(s)/D(s)$ is rational with non-repeated poles).

Using the impulse response matching method, the poles p_i of the transfer function $F_c(s)$ are mapped to poles $e^{p_i T}$ of $F_d(z)$ and

$$F_c(s) \text{ is rational of order } n \;\Rightarrow\; F_d(z) \text{ is rational of order } n$$

$$F_c(s) \text{ is asymptotically stable} \;\Rightarrow\; F_d(z) \text{ is asymptotically stable. (7.83)}$$

The zeros are mapped in a complicated way. In this case, however, the frequency responses of $F_d(z)$ and $F_c(s)$ can be related, provided the frequency response $\bar{F}_c(\omega) = 0$ for $|\omega| < \pi/T$, i.e., that $f_c(t)$ is a bandlimited signal sampled at twice its highest frequency. In that case, the results regarding the sampling of a continuous-time signal (7.27) indicate that

$$\bar{F}_d(\Omega) = \frac{1}{T} \left[\bar{F}_c(\omega) \right]_{\omega = \Omega/T} \tag{7.84}$$

for $|\omega| < \pi/T$, which is the desired result, *except for the factor of* $1/T$. For this reason, the impulse response procedure requires a slight modification of the formula (7.82), so that

$$F_c(s) \;=\; \frac{N(s)}{D(s)} = \sum_{i=1}^{n} \frac{c_i}{s - p_i}$$

$$\Leftrightarrow\; F_d(z) = T \sum_{i=1}^{n} c_i \frac{z}{z - e^{p_i T}} \quad \text{(impulse response equivalent).} \tag{7.85}$$

This approach is such that

$$f_d(k) = T \, f_c(kT), \tag{7.86}$$

where $f_d(k)$ is the impulse response of the discrete-time system, and $f_c(t)$ is the impulse response of the desired continuous-time system. One way to justify the factor T is to remark that a discrete-time impulse has an equivalent area T, while a continuous-time impulse has an area equal to 1.

Step response matching

The step response matching method is the same as the step response equivalent method that was used for the conversion from continuous-time to discrete-time system. The properties of the resulting transfer function are similar to those of the impulse response matching method, but the transfer functions are not exactly the same. For example, the step response method was shown to yield

$$F_c(s) = \frac{1}{s+1} \;\Rightarrow\; F_d(z) = \frac{1 - e^{-T}}{z - e^{-T}}, \tag{7.87}$$

while the impulse response method gives

$$F_c(s) = \frac{1}{s+1} \quad \Rightarrow \quad F_d(z) = \frac{T\,z}{z - e^{-T}}. \tag{7.88}$$

The poles of the transfer functions are identical, and the low frequency behavior (z close to 1) of both transfer functions is similar, but the transfer functions are different. Worth noting is the fact that the impulse response resulting from the step response matching method is delayed by one sample compared to the impulse response obtained with the impulse response method.

7.5.1 Sampled-data control design

The presentation of this section led to a conceptually opposite method to design sampled-data control systems, as compared to the one presented in section 7.3. The method of section 7.3 converted the continuous-time plant into an equivalent discrete-time plant, so that a discrete-time controller could be designed. The method of this section suggests a design of the controller in continuous-time, followed by an approximate implementation of the controller in discrete-time. The advantage of the first method is that implementation can be achieved with a direct realization of the controller in computer code. Some special algorithms, such as those based on FIR filters, are also specific to discrete-time. On the other hand, continuous-time design deals better with issues in the frequency domain, including robustness, and with physical systems having nonlinear dynamics.

In the early days of digital control, it appeared that discrete-time design would make continuous-time design obsolete. However, the high sampling rates and small quantization levels of modern systems have made it possible to implement continuous-time control systems with minimal sampling effects, even with crude Euler approximations. Thus, continuous-time design remains a common methodology despite the ultimate discrete-time implementation. Nevertheless, it remains important for the control engineer to have a good grasp of discrete-time and sampled-data issues, as they may significantly affect performance in some cases.

7.6   Appendix

7.6.1   Proof for the conversion from continuous-time to discrete-time signal

We first establish two facts. For an arbitrary function $f(t)$

$$f(t) = \int_{-\infty}^{\infty} f(t)\delta(t-\lambda)d\lambda = \int_{-\infty}^{\infty} f(\lambda)\delta(t-\lambda)d\lambda, \qquad (7.89)$$

where $\delta(t)$ is the delta function. Next, we note that the function

$$p(t) = \sum_{k=-\infty}^{\infty} \delta(t-kT) \qquad (7.90)$$

satisfies the following equality

$$p(t) = \frac{1}{T} \sum_{k=-\infty}^{\infty} e^{jk(2\pi/T)t}. \qquad (7.91)$$

Indeed, $p(t)$ is a periodic function, with period T. Its Fourier series is

$$p(t) = \sum_{k=-\infty}^{\infty} c_k \, e^{jk(2\pi/T)t}, \qquad (7.92)$$

with

$$c_k = \frac{1}{T} \int_{-T/2}^{T/2} p(t) \, e^{-jk(2\pi/T)t} dt = \frac{1}{T} \int_{-T/2}^{T/2} \delta(t) \, e^{-jk(2\pi/T)t} dt = \frac{1}{T}. \qquad (7.93)$$

Therefore, (7.91) is satisfied.

With these preliminaries, we now prove the result. The z-transform of the discrete-time signal is

$$X_d(z) = \sum_{k=0}^{\infty} x_d(k)z^{-k} = \sum_{k=-\infty}^{\infty} x(kT)u(kT)z^{-k}, \qquad (7.94)$$

where $u(t)$ is a continuous-time step signal. Since

$$\int_{-\infty}^{\infty} \delta(t-kT)dt = 1, \qquad (7.95)$$

(7.94) can be written as

$$X_d(z) = \sum_{k=-\infty}^{\infty} x(kT)u(kT) \; z^{-k} \int_{-\infty}^{\infty} \delta(t - kT)dt \tag{7.96}$$

and, since $\delta(t - kT) = 0$ unless $t = kT$,

$$X_d(z) = \sum_{k=-\infty}^{\infty} \int_{-\infty}^{\infty} x(t)u(t) \; z^{-t/T}\delta(t - kT)dt. \tag{7.97}$$

Permuting the order of integration and summation, and removing the step function in the expression by an adjustment of the integration range

$$\begin{aligned} X_d(z) &= \int_{-\infty}^{\infty} x(t)u(t) \; z^{-t/T} \left(\sum_{k=-\infty}^{\infty} \delta(t - kT) \right) dt \\ &= \int_{0}^{\infty} x(t) \; z^{-t/T} \left(\sum_{k=-\infty}^{\infty} \delta(t - kT) \right) dt. \end{aligned} \tag{7.98}$$

Next, using the preliminary fact (7.91)

$$X_d(z) = \int_{0}^{\infty} x(t) \; z^{-t/T} \left(\frac{1}{T} \sum_{k=-\infty}^{\infty} e^{jk(2\pi/T)t} \right) dt. \tag{7.99}$$

Permuting again the order of integration and summation

$$X_d(z) = \frac{1}{T} \sum_{k=-\infty}^{\infty} \left(\int_{0}^{\infty} x(t) z^{-t/T} e^{jk(2\pi/T)t} \right) dt, \tag{7.100}$$

one finds that

$$\begin{aligned} \left[X_d(z) \right]_{z=e^{sT}} &= \frac{1}{T} \sum_{k=-\infty}^{\infty} \int_{0}^{\infty} x(t) e^{-(s-jk(2\pi/T))t} dt \\ &= \frac{1}{T} \sum_{k=-\infty}^{\infty} X \left(s - jk\frac{2\pi}{T} \right). \end{aligned} \tag{7.101}$$

which establishes the result.

7.6.2   Proof for the conversion from discrete-time to continuous-time signal

The continuous-time signal can be written as

$$x(t) = \sum_{k=0}^{\infty} x_d(k) \left(u(t - kT) - u(t - (k+1)T) \right), \tag{7.102}$$

where $u(t)$ is the continuous-time step function. Applying the Laplace transform to both sides

$$X(s) = \int_0^\infty \sum_{k=0}^\infty x_d(k) \left(u(t-kT) - u\left(t-(k+1)T\right)\right) e^{-st} dt. \tag{7.103}$$

Permuting integration and summation

$$X(s) = \sum_{k=0}^\infty x_d(k) \int_0^\infty \left(u(t-kT) - u\left(t-(k+1)T\right)\right) e^{-st} dt. \tag{7.104}$$

Using the right shift formula of the Laplace transform and the expression for the transform of a step function

$$X(s) = \sum_{k=0}^\infty x_d(k) \left(\frac{e^{-skT} - e^{-s(k+1)T}}{s}\right). \tag{7.105}$$

Therefore, we have that

$$\begin{aligned} X(s) &= \left(\sum_{k=0}^\infty x_d(k) \left(e^{sT}\right)^{-k}\right) \frac{1-e^{-sT}}{s} \\ &= [X_d(z)]_{z=e^{sT}} \; \frac{1-e^{-sT}}{s}. \end{aligned} \tag{7.106}$$

which is the desired result.

7.7   Problems

Problem 7.1: (a) Consider the continuous-time system

$$H(s) = \frac{1}{s(s+1)}. \tag{7.107}$$

Find the discrete-time system $H_d(z)$ whose step response $y_d(k)$ is such that $y_d(k) = y(kT)$, where $y(t)$ is the step response of $H(s)$ and T is some arbitrary sampling time.
(b) Repeat part (a) for

$$H(s) = \frac{1}{s^2+1}. \tag{7.108}$$

Explain what happens when $T = 2\pi$.

Problem 7.2: Consider the signal

$$x(t) = 1 + \cos(20\pi t) + \sin(60\pi t). \tag{7.109}$$

Give the lowest sampling frequency f_s (in Hz) such that no aliasing occurs when the signal is discretized.

Problem 7.3: (a) The signal

$$x_d(k) = \cos\left(\frac{\pi}{2}k\right) \tag{7.110}$$

is sent to a D/A at a frequency of 1 kHz. Sketch the output waveform $x(t)$, making sure to label the time axis precisely.

(b) Using the discrete-time to continuous-time conversion results, sketch the magnitude of the Fourier transform of $x(t)$ in part (a). Give the frequencies (in Hz) and the magnitudes of the first three sinusoidal components, as well as the phase of the first component. Compare the results for the first component to those obtained by computing the coefficients of a Fourier series.

Problem 7.4: Let $x(t)$ be obtained from $x_d(k) = k$ through a zero-order hold. Find $X(s)$ from $X_d(z)$. Compare the result to the one obtained by computing the Laplace transform directly from $x(t)$ (note that $x(t)$ is the sum of step functions delayed by multiples of T).

Problem 7.5: A signal $x(t)$ with transform

$$X(s) = \frac{1}{s(s+1)^2} \tag{7.111}$$

is sampled at time instants $t = kT$ to obtain $x_d(k)$. Find the transform $X_d(z)$ of the resulting signal and obtain the poles.

Problem 7.6: Consider the continuous-time system

$$H(s) = \frac{2}{(s+1)(s+2)}. \tag{7.112}$$

Find the discrete-time system $H_d(z)$ whose step response $y_d(k)$ is such that $y_d(k) = y(kT)$, where $y(t)$ is the step response of $H(s)$ and T is some arbitrary sampling time. Compare the DC gains of $H(s)$ and $H_d(z)$.

Bibliography

[1] K. J. Astrom & R. M. Murray, *Feedback Systems: An Introduction for Scientists and Engineers,* Princeton University Press, 2008.

[2] R. W. Beard, T. W. McLain, & C. Peterson, *Introduction to Feedback Control Using Design Studies*, Amazon/independently published, 2019.

[3] S. Bennett, *A History of Control Engineering 1800~1930*, Institution of Electrical Engineers, London, 1979.

[4] M. Bodson, "Explaining the Routh-Hurwitz Criterion," *IEEE Control Systems*, vol. 40, no. 1, pp. 45-51, 2020.

[5] H. W. Bode, "Relations between Attenuation and Phase in Feedback Amplifier Design," *Bell System Technical J.*, vol. 19, no. 3, pp. 421-492, 1940.

[6] J. Chiasson, *An Introduction to System Modeling and Control*, Lulu Press, 2017.

[7] R. Clarke, J. J. Burken, J. T. Bosworth, & J. E. Bauer, *X-29 Flight Control System: Lessons Learned*, Technical Memorandum 4598, NASA, Washington, DC 20546, 1994.

[8] R. C. Dorf & R. H. Bishop, *Modern Control Systems*, 13th edition, Pearson, 2016.

[9] J. C. Doyle, B. A. Francis, & A. R. Tannenbaum, *Feedback Control Theory*, Dover, 2009.

[10] W. R. Evans, "Control System Synthesis by Root Locus Method," *AIEE Trans.*, vol. 69, pp. 66-69, 1950.

[11] G. F. Franklin, J. D. Powell, & A. Emami-Naeini, *Feedback Control of Dynamic Systems*, 7th edition, Pearson, 2014.

[12] G. F. Franklin, J. D. Powell, & M. Workman, *Digital Control of Dynamic Systems*, 3rd edition, Addison-Wesley, 1997.

[13] B. Friedland, *Control System Design: An Introduction to State-Space Methods*, Dover, 2005

[14] A. T. Fuller, "The Early Development of Control Theory," *J. of Dynamic Systems, Measurement, and Control*, pp. 109-118, 1976.

[15] A. T. Fuller, "The Early Development of Control Theory, II" *J. of Dynamic Systems, Measurement, and Control*, pp. 224-235, 1976.

[16] F. Golnaraghi & B. Kuo, *Automatic Control Systems*, 10th edition, McGraw-Hill, 2017.

[17] E. W. Kamen & B. S. Heck, *Fundamentals of Signals and Systems*, 2nd edition, Prentice-Hall, 2000.

[18] B. Kuo, *Digital Control Systems*, 2nd edition, Oxford University Press, 1995.

[19] W. S. Levine, *The Control Handbook*, 2nd edition, CRC Press, 2010.

[20] J. C. Maxwell, "On Governors," *Proc. of the Royal Society*, London, pp. 270-283, 1868 (available from: https://www.jstor.org/stable/112510).

[21] O. Mayr, *The Origins of Feedback Control*, MIT Press, Cambridge, MA, 1970.

[22] N. S. Nise, *Control Systems Engineering*, 7th edition, Wiley, 2014.

[23] H. Nyquist, "Regeneration Theory," *Bell System Technical J.*, vol. 11, pp. 126-147, 1932.

[24] K. Ogata, *Discrete-Time Control Systems*, 2nd edition, Pearson, 1995.

[25] K. Ogata, *Modern Control Engineering*, 5th edition, Pearson, 2009.

[26] A. V. Oppenheim, A. S. Willsky, *Signals & Systems*, 2nd edition with S.H. Nawab, Pearson, 1996.

[27] C. L. Phillips & R. D. Harbor, *Feedback Control Systems*, 4th edition, Prentice-Hall, 2000.

[28] J. R. Ragazzini & L. A. Zadeh, "The Analysis of Sampled-Data Systems," *AIEE Trans.*, vol. 71, pp. 225-234, 1952.

[29] E. J. Routh, *A Treatise on the Stability of a Given State of Motion, Particularly Steady Motion*, Macmillan & Co., London, 1877.

Index